EARTHLY PARADISES

EARTHLY PARADISES

ANCIENT GARDENS IN HISTORY AND ARCHAEOLOGY

MAUREEN CARROLL

THE BRITISH MUSEUM PRESS

This book is dedicated to my parents

First published in 2003 by The British Museum Press
A division of The British Museum Company Ltd
46 Bloomsbury Street, London WC1B 3QQ

Reprinted in 2004

A catalogue record for this book is available from the British Library

ISBN 0 7141 2768 X

Designed by Andrew Shoolbred
Typeset in Garamond Light
Printed in Spain by Grafos SA, Barcelona

Frontispiece *The Garden of Eden is depicted with stylized trees and the four rivers of Paradise in this scene from the illustrated Byzantine Homilies of Jacob Kokkinobaphos, twelfth century* AD.

Contents

Preface

The gardens of some ancient cultures are as legendary today as they were in antiquity. The hanging gardens of Babylon, for example, were considered centuries later by the Greeks to be one of the seven wonders of the ancient world. The royal gardens of the ancient Persian kings were admired by Greeks, Romans and Arabs alike, and the garden of paradise at the beginning of time, the biblical garden of Eden, was sometimes compared by Byzantine writers to the contemporary palace gardens of the emperors.

Looking at historic gardens has become increasingly popular, especially in the last thirty years. Archaeological excavations have contributed much to our understanding of parks and gardens and, with all the recently conducted work on gardens of the post-medieval period, it has been declared that garden archaeology is 'the latest fashion in British archaeology' (*Current Archaeology* 12, 292, 1994). Archaeology has also increased our awareness of far older gardens, which were cultivated millennia ago in many cultures of the ancient world. While earlier publications dealing specifically with ancient gardens frequently relied heavily on literary and archival sources, the unique contribution of archaeology is becoming increasingly clear. Excavations can reveal the remains of ancient gardens that were never mentioned in written sources, particularly those that did not belong to the rich and famous.

Although most of the archaeology concerning the gardens of antiquity has taken place in Italy, in particular in Pompeii, Herculaneum and the Roman cities and farms buried by the eruption of Vesuvius in AD 79, in the last decade there have been archaeological investigations of Roman gardens elsewhere, including Portugal, France, Switzerland and North Africa. In Britain, the spectacular results of the excavations in the Roman palace garden at Fishbourne in

1962 have not been duplicated at any other Roman villa in the country, but evidence for other gardens is gradually being unearthed on a smaller scale at other sites. Recent studies on the gardens in ancient Greece and Egypt have also contributed significantly to a better understanding of gardening and gardens in these ancient cultures.

My own interest in ancient gardens began in 1989 with an investigation of Classical and Hellenistic Greek settlements, and what role gardens played in them, and in 1992 led to a multi-phased study of gardens from ancient Egypt to medieval Europe. The results of both investigations were published in German, the latter having gone into a third printing in 1998. Given the general interest in ancient gardens, and the popularity of gardens and gardening in the media, I have produced this up-to-date study of archaeological garden excavation with a broader scope than ever before.

This book mainly spans the period from the second millennium BC to the middle of the first millennium AD, and it examines the material evidence for gardens in the cultures of the ancient Near East, Egypt, Greece, Italy and the provinces of the Roman Empire. In the final chapter, the survival of ancient gardening traditions in the Islamic and Byzantine worlds is explored, as is the perception and depiction of paradise in those cultures. Throughout the book, themes such as house gardens, orchards and parks, sacred gardens and cemetery gardens are examined, and the roles of gardens in society, the economy and religion are discussed. The book also looks at the tasks and role of gardeners and assesses the evidence for various types of plants, trees and horticultural practices. Archaeological, textual, pictorial and environmental evidence has been integrated to gain a fuller picture of ancient gardens. Only in a synthesis of all available sources of information can we hope to catch glimpses of the earthly paradises of antiquity.

I am grateful to the British Museum Press for taking up the idea for this book, and for the careful supervision of its production by Laura Brockbank. Many thanks are due also to the various authors, institutions, publishers and illustrators who have provided photographs and drawings, and allowed them to be reproduced.

CHAPTER ONE

Ancient gardens and the evidence

Gardens are ephemeral, as fragile and short-lived as the people who once cultivated them. They are closely associated with the values of the societies in which they are created. In all ancient cultures of the Mediterranean and Europe, gardens were essential as sources of food but, on another level, they were a sign of affluence and prestige, particularly the ornamental and work-intensive gardens associated with the villas and palaces of the wealthy and powerful. In many cases they were also directly associated with the gods and their divine powers.

While the physical remains of ancient plantings and vegetation, such as roots, leaves, stems, seeds and fruit, only occasionally survive under favourable conditions, there are many other types of evidence for gardens of both the prehistoric and historic periods. These range from actual garden features such as planting beds, trenches and garden furniture to written records and pictorial representations in paintings and sculpture.

The available sources on ancient gardens reveal that some cultures had a recognizably grand tradition of gardening. In Egypt, for example, the garden not only played an important role in the agrarian economy supplying the population with numerous essential products, but it was also associated with luxury and leisure as well as divine blessing. Just as the Egyptians imagined the gods to inhabit the gardens and groves of their temples, so the pharaohs, as the divine rulers on earth, enjoyed extensive gardens in and around their palaces. In the next – and better – life, ancient Egyptians hoped to partake of the pleasures and shaded peace of gardens, which are consequently often represented in tomb paintings and in scenes in the *Book of the Dead* (fig. 1), a compilation of spells pertaining to death and the Underworld. The fertile silt deposits left by

1 *The deceased and his wife in their garden,* Book of the Dead *papyrus of Nakht,* c. *1300* BC.

the annual floods along the banks of the Nile and at the delta were used in the cultivation of gardens and, despite the encroaching desert sands and aridity of the rest of the country, gardens, pools and lakes were maintained at many sites by the development of effective irrigation systems.

In Roman Italy, there was hardly any aspect of daily life that did not relate to gardens in one way or another. The archaeological investigations of the cities and farms buried by the volcanic eruption of Vesuvius in AD 79 have shown that almost all private houses had at least one garden (fig. 2). It could dominate the house and occupy a surface area larger than the rooms used for sleeping, dining and socializing. Even if space was limited in a very small house, a corner of the courtyard was set aside for a modest planted area. For those who lived in blocks of flats, window boxes could provide space for growing plants, although they could be so small that 'a cucumber [could not] lie straight in it and a snake [could not] live in it at full length' (Martial, *Epigrams* 11.18). Gardens and groves surrounded the temples in Roman towns as well as

2 View through the atrium to the courtyard garden in the House of the Vettii, Pompeii, first century AD.

forming part of taverns and dining establishments, where the customer could enjoy his sojourn in a leafy environment.

The location of gardens and orchards was determined by the availability of water. A sophisticated system of aqueducts supplied Roman towns with enough water not only for drinking, cooking and washing but also for irrigating gardens. Those cultures that did not build aqueducts or artificially pipe water into their towns, settlements and houses depended on natural sources. Mesopotamian settlements were often surrounded by a green belt of groves and orchards, particularly groves of date palms, where a nearby river made cultivation possible (fig. 3). In Classical Greece, there was a relatively clear separation between town and country, between the space inhabited and built by people and the less crowded suburbs and countryside where vegetation flourished. Greek towns were surrounded by a green belt of vegetation, by utilitarian garden plots and suburban groves, all located near rivers outside the towns. If any planted and irrigated areas did exist in Greek towns, they were public

3 Assyrian relief from Nineveh showing the Elamite city of Madaktu surrounded by palm groves on a river, c. 650 BC.

ones. Supplying urban areas with water was a problem in ancient Greece and it was not until the Roman period that private homes were connected to city water pipes, allowing the maintenance of a house garden.

Natural resources were very important to all cultures of the ancient Mediterranean and Europe, which despite the existence of trade and industry, were primarily agrarian societies. Human contact with nature was close and intimate, and a great deal of time and effort was invested in cultivating gardens, groves and fields. An ancient Egyptian text dating to around 1800 BC describes the work of a gardener: 'The gardener carries the yoke; his shoulders are bent as with age. There's a swelling on his neck, and it festers. In the morning he waters the vegetables, the evening he spends with the herbs, while at noon he has toiled in the orchard. He works himself to death, more than all other professions' (*Instruction of Khety*). It is no wonder that, in Egyptian paintings and reliefs, agricultural labourers are occasionally depicted resting from the fierce heat in the shade of a tree (fig. 4).

4 An agricultural labourer seeks shade and drinks from a waterskin under a tree on an Egyptian painting in the tomb of Nakht at Thebes, c. 1390 BC.

5 Four fruit trees and pairs of animals alluding to paradise on a floor mosaic in the Church of Saints Lot and Procopius on Mount Nebo in Jordan, c. AD 550.

Perhaps because of the strenuous efforts required to maintain a garden, many cultures shared the perception of a flourishing garden as a blessed place in which hard physical labour, wind and weather played no role. In this respect, the Egyptian conception of the world of their gods as a fertile garden had much in common with the Christian and Islamic perception of paradise as a garden; in both cases, paradise was that place in which peace prevailed in a fertile, green and well-watered environment (fig. 5). This paradisiacal and ideal world, the garden of Eden, can be linked to the Near East, to the verdant palm groves and pleasure parks on the Tigris and Euphrates rivers. Pagan perceptions of a type of paradise, or Elysium, for gods and heroes also existed in Greek mythology from at least the late Bronze Age. In a mythical grove near the Atlas Mountains, the so-called garden of the Hesperides, the Greeks believed that the goddess Hera planted the tree of immortality that was laden with golden apples. In this garden grew a variety of other trees such as pomegranate, pear, mulberry, myrtle, laurel and almond (Hesiod, *Theogony* 215–16; Skylax, *Periplus* 108). In both the Greek and Roman world, funerary monuments were often erected within enclosed gardens reflecting the desire that both the souls of the dead and their visitors should find pleasure in the garden at the tomb.

This brief introduction shows why it is so important to study gardens in any past culture – after all, the cultivation of plants and the maintenance of gardens and groves played an integral role in public, private, economic and religious spheres everywhere. Interest on the part of historians and art historians in ancient gardens is not a recent phenomenon, although the archaeological investigation of them is a relatively new development. If we look at just one case study, that of Roman Pompeii, archaeologists in the nineteenth century and for most of the twentieth century were more concerned with uncovering

structural remains such as walls, floor mosaics, wall paintings and the moveable contents of private and public buildings. It was not until the 1970s that research was focused on the 'empty spaces' in public places, next to houses and within courtyards, with the intention of examining the subsoil for remains of gardens. As excavations progressed in Pompeii, it became clear that these 'empty spaces' were not empty at all, they were in fact landscaped and designed to be an integral part of the structural complex.

6 Roman garden designed for the display of Roman sculptures from the Mahdia shipwreck in the Rheinisches Landesmuseum in Bonn.

Excavating, recording and analysing this material has given far deeper insights into the daily lives of the people of Pompeii than any textual sources on the city have ever done. It has helped develop our understanding of the overall design and complexity of a Roman house and increased our appreciation of the garden as an environment in its own right. As a result of this shift in attitude, I was asked in 1994 by the Rheinisches Landesmuseum in Bonn to design a Roman garden as the setting for an exhibition of the sculptures retrieved in the cargo of a shipwreck off the Tunisian coast near Mahdia (fig. 6).[1] These sculptures were destined for Roman villa gardens in Italy when the ship carrying them sank around 80 BC, and the aim was to show them in their 'proper' context, rather than in a sterile gallery environment.

The American historian and archaeologist, Wilhelmina Jashemski, pioneered the discipline of garden archaeology in the Vesuvian cities of Italy, systematically re-examining open spaces that had been cleared of the volcanic debris of AD 79, some of which had been excavated long before she began her research in 1955.[2] She examined the surface of these areas, detecting the contours in the soil of planting beds, irrigation channels, the borders of flowerbeds and even the outlines of a wooden ladder once used to pick cherries from a tree in a courtyard before the eruption of Vesuvius. She also realized that the roots of the trees, bushes and shrubs that had been growing at the time of the disaster had slowly decomposed, leaving voids in the subsoil which gradually filled up with small pumice pebbles (*lapilli*) from the deposits of the same

7 Cement cast of an ancient tree buried in the eruption of Vesuvius in AD 79 at the Roman farm at Boscoreale near Pompeii.

material on top of the soil. By clearing the *lapilli* from the root cavities and filling the emptied cavities with liquid plaster or cement, a cast of the roots, or sometimes of even an almost complete tree trunk, emerged which could be analysed and identified by botanists (fig. 7). This method had been applied earlier to produce casts of the bodies of Pompeians who were buried under pumice and ash. It has been employed ever since in excavations of gardens in Pompeii and because of its application we now know that private gardens, orchards, vineyards and plant nurseries were maintained throughout the city.

In 1998 I opened trenches in the courtyard of the temple of Venus in Pompeii, a temple that was originally excavated in the late nineteenth century. At that time the temple's courtyard had been ignored, but during the re-excavation we found evidence of a planting pit, probably for a large tree, and a smaller pit on one side of the temple that contained an intact ceramic planting pot. The latter would once have contained the root-ball of a shrub and this shrub can only have been one of a series, planted to form a hedge. Closer to home, in Britain, excavations have been conducted since 2000 by a team from the University of Sheffield, including myself, at Chedworth Roman villa in Gloucestershire. The structural remains of the villa were uncovered in the late nineteenth century but, as our trenches showed, the Victorian excavators had not investigated the large courtyard in the middle of the ranges of rooms – probably because it was not as appealing as other treasures such as the mosaics. After only three seasons of re-excavation, it is clear that the courtyard was laid out as a garden in the early fourth century AD, with large amounts of soil having been dumped on earlier, abandoned structures to raise the level of the courtyard and transform it into an important part of the villa complex.

Of course, not all ancient sites were buried, as Pompeii, under volcanic ash which provided such ideal conditions for the preservation of plantings, nor are all excavations conducted as long-term research projects. Most excavations, at least in Europe, are rescue digs conducted immediately before the construction of a new building. Time and money are limited and work is usually concentrated on more tangible, easily recognizable structures. Furthermore, many Roman sites are buried deep beneath post-Roman, medieval and modern deposits and buildings, and the Roman features may have been heavily disturbed in later periods. However, whenever possible it is important to pay

8 Assyrian relief from the palace of Ashurbanipal in Nineveh showing groves, orchards and water courses outside the city, c. *650* BC.

attention to open spaces associated with Roman buildings as they were an essential part of the overall design, even if the chances of finding intact soil deposits, in what once may have been a courtyard garden, are fairly remote.

Each scrap of evidence for ancient gardens has its own value and may contribute to further understanding. Pictorial sources can be extremely useful: a seventh-century-BC stone relief from the palace of Ashurbanipal at Nineveh, for example, depicts a park outside that city and informs us how such parks and groves were irrigated by stone aqueducts and canals (fig. 8). An Egyptian tomb painting in Thebes depicts the deceased Sebekhotep and his wife in the midst of a garden planted with palms and sycomore trees, and at the centre of this garden is a pool with fish and lotus flowers (fig. 9). The painting indicates the type of vegetation that grew in Egyptian gardens and how carefully and symmetrically planned these gardens were, at least theoretically.

9 Sebekhotep and his wife in their garden of sycomores, dom palms and date palms, tomb of Sebekhotep, c. *1550–1295* BC.

Written sources also contain information on gardens. An inscription of the late fourth century BC from the Greek island of Thasos tells us that a garden

10 Rows of rock-cut planting pits for trees or shrubs around the temple of Hephaistos in Athens, c. *third–first century* BC.

of myrtle, fig and nut trees belonging to the temple of Herakles was leased to a private individual.[3] An Alexandrian papyrus, dating to around 5 BC, records tomb gardens planted with cabbage, asparagus, leeks, grape vines and date palms outside the city and informs us that these were contracted out to individuals for a period of five years at a cost of twenty drachmas a month.[4] Both documents provide evidence for gardens as a source of income for the owners and also give us an indication of the type of plantings that they contained. Pliny the Younger's loving description of his Tuscan villa in Italy in the first century AD, particularly his lengthy references to terraces, trees, hedges, meadows and kitchen gardens, are eloquent testimony to the high esteem in which Romans held their gardens (*Letters* 5).

Excavated archaeological evidence for gardens can come in the form of rows of planting pits for individual trees. There are perhaps in total about two thousand of these in and around the temple of Ashur, built by King Sennacherib (705–681 BC) on the Tigris river in Assyria.[5] Evidence can also take the form of ceramic planting pots such as those found in rows of pits running parallel to the walls of the temple of Hephaistos on the edge of the Athenian agora (fig. 10).[6] These plantings, possibly dating to the third century BC, may have consisted of myrtle and pomegranate bushes. Furthermore, excavated evidence, such as the outlines of decayed tree roots in the courtyard of the house of Ariadne and Bacchus at Thuburbo Maius, allow the reconstruction of a small, informally planted domestic garden in Roman Tunisia.[7]

Environmental evidence is also increasingly important in garden archaeology. Leaves and clippings of box, for example, have been found at many sites, including several Roman sites in Britain, such as Winterton villa in Lincolnshire.[8] Carbonized grapes, olives, apples and cherries, show some of the types of trees grown in Pompeii. Likewise at Pompeii, clumps of olive pollen in the soil of a number of gardens, sometimes up to 94 per cent of the total pollen, attest to the presence of olive trees in this environment.[9] Furthermore, imported topsoil, enriched with compost and brought into a garden site to improve the cultivation of plants, has been found in a number of places, including the fourth-century AD Roman villa at Montmaurin in France.[10] These are but some of the types of written, pictorial, archaeological and environmental evidence that help to complete the picture of ancient gardens, even if still very imperfectly.

Our modern term 'garden' is a very general one and often further qualification is needed to clarify exactly what type of garden is being described – whether a flower garden, a vegetable garden or a herb garden. Likewise, gardens in the ancient world were varied, and different words were used to describe them. In ancient Greek, the generic term for a garden was κῆπος (*kepos*), while that for a tree garden or orchard was ἄλσος (*alsos*), and ἀμπελῖτης γῆ for a vineyard. To confuse the issue, however, *kepos* could mean plot of land cultivated solely or chiefly for trees. However, in ancient Egypt, a single word, *k3mw* or *k3nw*, appears to have referred equally to a garden, an orchard or a vineyard. The word for 'garden' and 'orchard' is the same (*kiru*) in ancient Akkadian, but evidence suggests that the term *kirimahu*, meaning 'pleasure garden', was introduced by Sargon II in the late eighth century BC.

In each case, the plants grown on a plot of land need to be defined in order to understand the character of the garden. In Latin, the division is clearer between a tree-planted area or a grove (*nemus*) and a garden cultivated with other plants (*hortus*), although even the large, luxury estates with gardens containing all manner of trees, plants and statuary in early Imperial Rome were sometimes referred to as *horti*. Given the diversity in the appearance, role and location of ancient gardens, the context of the garden and its plantings is always of importance in attempting to understand its function.

We can see, then, that the evidence for ancient gardens is diverse, and it varies in its nature in each country, time period and context. In the following chapters, we will be exploring the available and disparate sources of information on gardens in an attempt to gain a fuller picture of horticulture in the many cultures of the ancient world.

CHAPTER TWO

Utilitarian and ornamental house gardens

In some ancient cultures it was common to own a house and a garden. 'Field, garden and house' as a unit appears frequently as a reference in legal and administrative texts in Mesopotamia, and in Greek inscriptions from the fifth to the first century BC, 'house and garden' is a common entry in records of real-estate property. The garden was often quite separate from the house and the two could be bought and sold separately.

Gardens could provide the owner with fruit and vegetables, either for his own consumption or for sale. Like many other country dwellers, the mother of the Athenian playwright, Euripides, is said to have raised vegetables commercially on her farm outside Athens for markets in the city in the fifth century BC (Aristophanes, *Thesmophoriazusae* 387). The Roman author Cato, in his second-century treatise *On Agriculture*, wrote that the suburban garden ideally should be planted with useful plants such as flowers for wreaths, onions, myrtle for wedding celebrations, laurel and nut trees (*De Agri Cultura* 8.2).

Utilitarian gardens depicted in Egyptian wall paintings consisted of plots of land subdivided into square beds and, depending on the available space, they could be laid out on a grand scale or fitted into smaller plots between the houses. On a relief in the tomb of the official Mereruka at Saqqara dating from around 2330 BC, garden labourers irrigate gardens divided up into square beds and crossed by canals (fig. 11).

Archaeological evidence for regular rows of square planting beds has been uncovered at several sites, including Mirgissa in Sudan, where the edges of the individual plots were raised so that the water with which the plants were irrigated would sink into the small depression rather than run off or evaporate quickly (fig. 12).[1] In Egypt at the Workmen's Village at Amarna, a city established

Detail of fig. 16 (p. 27)
The garden of King Ashurbanibal on a relief from Nineveh, c. 650 BC.

11 Egyptian labourers watering plants in a garden subdivided into plots, tomb of Mereruka at Saqqara, c. 2330 BC.

12 Square planting beds in an excavated Egyptian garden at Mirgissa in Sudan, Middle Kingdom and Second Intermediate Period, c. 2125–1550 BC.

by the pharaoh Akhenaten around 1350 BC, walled areas adjacent to the houses contained square growing plots separated by low mud-brick partitions, each plot containing a deposit of alluvial soil.[2] The dimensions of the plots or plant receptacles at Mirgissa and Amarna varied only slightly, suggesting that there may have been a relatively standard size of growing plot in such regular layouts.

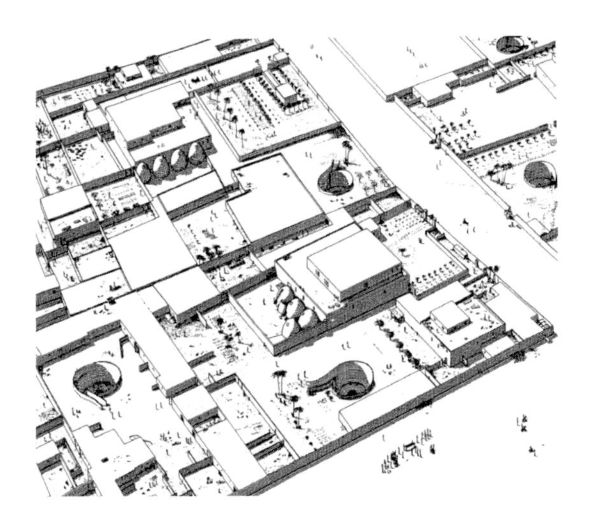

13 Reconstruction of Egyptian houses and planted courtyards in Amarna, c. 1350–1334 BC.

Although it is not known what plants grew in these beds, a variety of vegetables must have been cultivated. While Egyptian pictorial sources repeatedly represent only lettuce and onions, archaeological evidence shows that vegetables such as leeks, lettuce, cucumbers, radishes, lentils and beans were grown in the second millennium BC in Egypt and lower Mesopotamia, areas of dense settlement and arid conditions. Because of the bordering deserts in both regions, the use of limited areas of irrigation or flooded land was essential to meet the needs of the population, especially since vegetable crops had to compensate for the shortage of wild plants in the desert areas.[3]

If the house was large enough, an orchard would be found near the Egyptian vegetable garden. The excavated villas of high administrative officials and other houses in the main city at Amarna, ancient Akhetaten, had relatively large plots of open land surrounded by an enclosure wall, and judging by the surviving planting pits in the soil, the land was used for the cultivation of gardens and orchards (fig. 13).[4] Within the enclosure there were also storage silos, wells, animal byres and ovens. Such houses, built by court officials and overseers, are recorded in autobiographies carved on the walls of their tombs. In his autobiography, Harkhuf, governor of Upper Egypt, recorded for posterity: 'I have come here from my city, I have descended from my nome; I have built a house, set up [its] doors, I have dug a pool, planted sycomores.'[5] A scribal school text, recording a house in a town and a farm in another village, reveals that some people possessed property in more than one location: 'I will build for you a new villa upon the ground of your city, surrounded with trees . . . I will tend for you five arouras of cucumber beds to the south of your village, and the cucumbers, carobs and . . . will be as abundant as the sand' (*Papyrus Anastasi* IV).

Egyptian paintings portray domestic orchards enclosed by a wall and full of the types of trees that were cultivated most frequently, ranging from the sycomore fig (*Ficus sycomorus*) and the common fig (*Ficus carica*) to the date palm and the pomegranate. Paintings from tombs reveal such orchards were positioned near to private houses and domed storage silos. This close proximity is also confirmed by a small coniferous wood model from the tomb of Meketre at Thebes revealing a house with a garden of sycomore trees around a pool (fig. 14).

The pleasure gardens of ancient Egypt are most frequently associated with the estates of the wealthy, particularly those of court officials and of the pharaohs themselves. Typically they are represented in art surrounded by a wall and located next to a large, multi-roomed house, or they are in a courtyard of the house and the garden is often associated with a pool in which lotus flowers grow and fish swim. Flower-beds were frequently planted around the edges of pools, too. Some of the possible varieties of vegetation in a pleasure garden of this type are recorded in the Theban tomb of Ineni of the sixteenth century BC: these include sycomores, fig trees, date palms, dom palms (*Hyphaene thebaica*), persea (*Mimusops laurifolia*), olive trees, pomegranates, willows, tamarisks and grape vines.[6]

But pleasure gardens are not only known through their depiction in Egyptian art. Excavations in 1979–89 at Tell el-Dabaa (ancient Avaris) in the eastern Nile delta uncovered a palace of the Thirteenth Dynasty (eighteenth century BC) surrounded by surprisingly well-preserved remains of gardens.[7] To the north, south and east of the palace were gardens with square flower-beds set in a regular grid pattern and trees arranged around an (unfinished) pool. The planting pits for the trees were encased in circular brick borders and were supplied with water from irrigation ditches. Judging by the substructure of a possible wine-press, another garden at the site may have been planted with vines. Although the settlers at Tell el-Dabaa were Canaanites, possibly living there as traders under Egyptian jurisdiction, the palace is purely Egyptian in its layout and the gardens also conform to those at other Egyptian sites.

14 Painted wooden model of an Egyptian house and garden with sycomore trees from the tomb of Meketre, c. *1990 BC.*

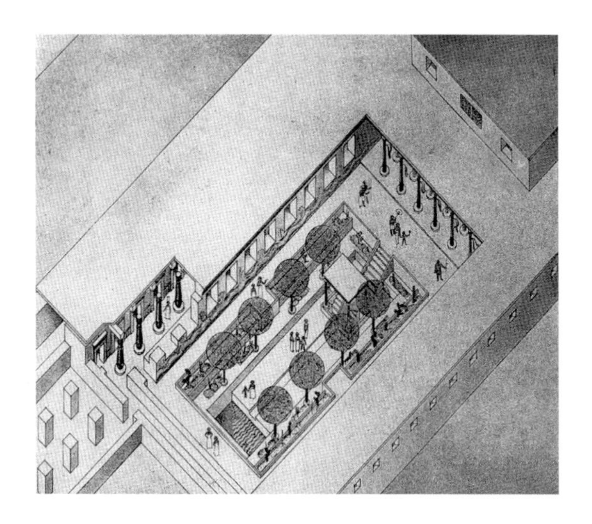

15 Reconstruction of the so-called harem and its courtyard garden at Amarna, c. *1350–1334* BC.

The pleasure gardens of the pharaoh are best known at Amarna. In the northern section (the so-called harem) of the royal palace a courtyard was laid out on several terraces which were planted with trees and flower-beds, and irrigated by a well, a channel and a pool (fig. 15). In the southern section of the city at Amarna stood a palace complex, the Maru-Aten, that once belonged to Akhenaten's eldest daughter Meritaten. Artificial lakes, the largest one measuring 60 × 130 metres in size, numerous pools, flower-beds, arbours and trees adorned the open areas within the complex.[8] Gardens such as these were designed and understood as overt symbols of status and power. They were also extremely labour-intensive, particularly in the dry climate of Egypt. Digging the wells, pools and lakes; bringing up the water through channels; watering and fertilizing the plants and building and maintaining both the buildings and the gardens on such sizeable parcels of land needed a large entourage of personnel.

The Egyptian pleasure garden was also a place associated with romantic encounters, and song and sensual pleasure. Egyptian love poetry makes frequent reference to the garden; in a poem of the Nineteenth Dynasty a lady sings: 'I am your best beloved. I am yours like this field which I have planted with flowers . . . a lovely place for strolling, with your hand in mine. My body is satisfied, my heart in joy at our going out together. Hearing your voice is pomegranate wine; I live for hearing it, and every glance which rests on me means more to me than eating and drinking' (see Chapter 8).[9]

That the pleasure garden was also a place for rest, relaxation and enjoyment in the first millennium BC in Mesopotamia is indicated by an Assyrian relief depicting King Ashurbanipal (668–626 BC) dining, drinking and chatting under a vine trellis in a royal garden planted with various types of trees (fig. 16). Perhaps the most famous Near Eastern pleasure gardens in antiquity – and the ones we know the least about today – were the so-called 'hanging gardens' of Babylon.[10] This garden, almost certainly connected to the palace, was constructed between 604 and 562 BC by Nebuchadnezzar II for his wife,

16 The Assyrian King Ashurbanipal dines in his royal garden on a relief from Nineveh, c. 650 BC. *Birds perch in the trees and the severed head of the defeated Elamite ruler hangs from a branch.*

Amytis, who was homesick for her native Media (Kurdistan) and the mountain vegetation with which she was familiar. It was considered one of the ancient world's seven wonders but has not been located with any certainty in excavations of the palace. The only details we have are those supplied by Greek and Roman authors, who describe a terraced garden approximately 120 metres long and 25 metres high, supplied with water from the nearby Euphrates river. In general, the pleasure garden most frequently encountered in the ancient Near East is the orchard, grove or park (see Chapter 3). However, a royal herb and vegetable garden belonging to Merodach-Baladan II, who ruled from Babylon in the late ninth century, is also mentioned in a cuneiform text.[11] Presumably vegetable plots and herbs, possibly for medicinal use, were cultivated in more modest gardens.

In ancient Palestine, an arid land, gardens were highly prized if we are to believe the tale of King Ahab of Samaria who coveted the vineyard of Naboth and wanted to transform it into a 'garden of herbs' so that it would be an asset to the palace (First Book of Kings 21.2). Pleasure gardens are a recurring topic in the Song of Solomon: 'A garden enclosed is my sister, my spouse; a spring shut up, a fountain sealed. Thy plants are an orchard of pomegranate, with pleasant fruits; camphire, with spikenard, and saffron; calamus and cinnamon, with all trees of frankincense; myrrh and aloes, with all the chief spices . . .' (4.12–15). This unabashed appreciation of gardens, however, is rarely expressed in biblical verses of the Old Testament; more common is the idea that the pleasure given by gardens and orchards is mere vanity: 'I builded me houses, I planted me vineyards; I made me gardens and orchards, and I planted trees in them of all kind of fruits; I made me pools of water, to water therewith the wood that bringeth forth trees . . . and behold all was vanity and vexation of spirit, and there was no profit under the sun' (Ecclesiastes 2.4–6, 11). Nevertheless, despite the moralistic condemnation of the sensuous pleasures of the garden, biblical references clearly confirm the existence of well-watered, fertile gardens.

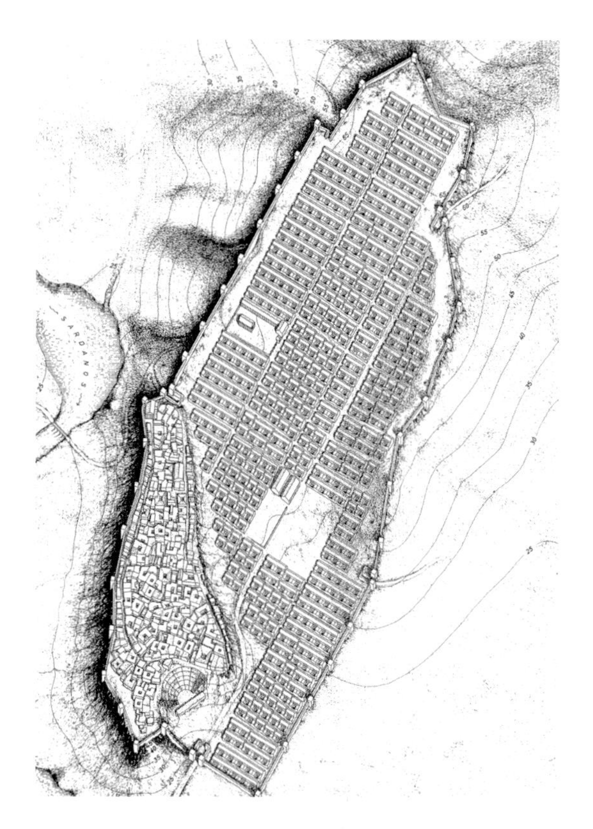

17 Bird's-eye view of the Greek city of Olynthos after 432 BC.

Precious little is known about either utilitarian or ornamental gardens in the Greek world before the eighth century BC, and even then the source of information is purely literary.[12] In Book VII of Homer's *Odyssey*, at the palace estate of Alkinoos, the Phaeacian king is described as having a garden with three main elements: an orchard with apple trees, pear trees, olive and pomegranate trees; a vineyard; and a vegetable garden. Judging by this description, these were not royal pleasure gardens, but rather well-ordered utilitarian gardens. There is considerably more evidence for gardens of various kinds from the fifth century BC in Classical Greece and the eastern Mediterranean.[13] Excavations in all parts of the ancient Greek world have shown that the city within the defensive walls was densely built up, and that its residents lived at close quarters. Whether in irregularly laid-out cities, such as Athens, or in towns laid out on a regular grid plan, such as Olynthos or Priene, the individual houses were built adjacent to each other and they bordered the street (fig. 17). The average building plot was only 250 square metres, and every bit of it was necessary for the house itself.

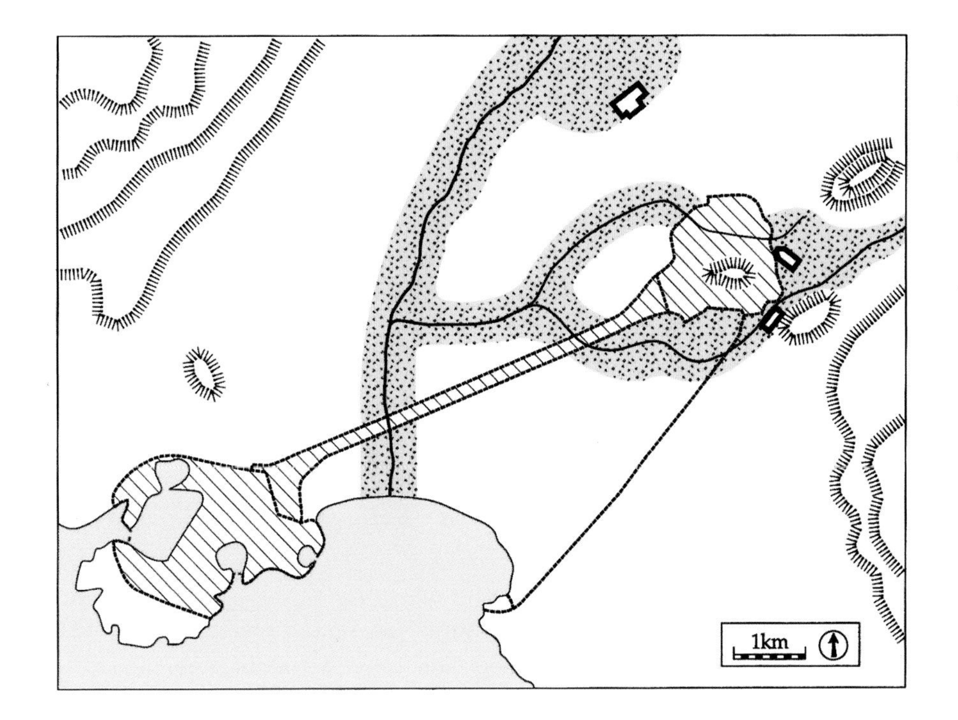

18 Classical Athens and its suburbs with garden zones along the rivers (dotted). The three gymnasia (see p. 50) in the suburbs are outlined in black, and the walled city, including the corridor to the harbour at Piraeus, is cross-hatched.

If there were no cisterns in the city itself, and if there were no springs, streams or wet areas within the city walls, the only suitable location for gardens was beyond the city where rivers and streams could supply the gardens with water. In Classical Greece gardens were clustered around the city in a green belt of vegetation in suburbs and rural districts (fig. 18). Many of these were market gardens for fruit, vegetables and flowers, whereas fields of grain, vineyards and orchards were more prevalent on the outlying farms. Such gardens are referred to in literary works and in inscriptions recording real-estate property, but in very few cases is the actual physical relationship between the house and the garden specified. The utilitarian garden on the farm was set apart from the house and protected from human and animal intruders by a wall.

It appears that gardens were a relative luxury in Classical Greece. Thanks to the survival of numerous inscriptions and literary sources, we are informed about the purchase prices for gardens in the fourth century BC, primarily in Athens and its environs.[14] Plato bought a garden in the suburbs of Athens in 388 BC for 2,000 drachmas, and Epicurus purchased one in the same century for 8,000 drachmas. Further away from the city, in Attica, a garden located inland cost 250 drachmas, whilst a house and a garden in the coastal region was sold for only 205 drachmas. Since the average daily wage of a labourer in Classical Athens was only one drachma and anyone who had a personal wealth of less than 2,000 drachmas at the end of the fourth century BC was not included in the propertied classes, gardens as costly as those purchased by Plato and Epicurus would have been the preserve of the rich. But gardens could be leased for between thirty and seventy drachmas a year in Athens, so that some of the poorer residents would at least have had access to one. Perikles, speaking after the destructive events of the war against Sparta in the late fifth century, encouraged the Athenians not to mourn the loss of possessions that were but mere luxuries – and revealingly one of the luxuries he mentioned was a garden (Thucydides, *Peloponnesian War* 2.62).

By the Hellenistic period, that is roughly the third to first centuries BC, the conditions for house gardens within the Greek city had not substantially improved and houses were still crowded together. Building laws of Alexandria, for example, specified that a distance of 30 centimetres between houses had to be maintained. Internal courtyards, surrounded by columned walkways (peristyles), were still regularly paved, many of them with mosaics. The suburbs remained the prime location for utilitarian gardens, particularly in Alexandria, the city founded in the late fourth century BC by Alexander the Great on the coast of northern Egypt, where suburban gardens and tomb gardens were planted with fruit trees and vegetables (fig. 19).

It has often been assumed that villas of wealthy Hellenistic Greeks had pleasure gardens similar to those of the later Roman villas.[15] However, even if some urban houses in Hellenistic cities could be much larger than their Classical predecessors, there was still no available space for a garden, and courtyards still tended to be paved. Only in Alexandria did the palaces built by the Greek King Ptolemy and the later kings of his dynasty appear to have planted areas. Strabo, writing his *Geography* (17.1.8) in the early first century AD, described the peninsula of land leading to the harbour of Alexandria as the location of the palaces, groves and parks. These palace parks are almost certainly a traditional

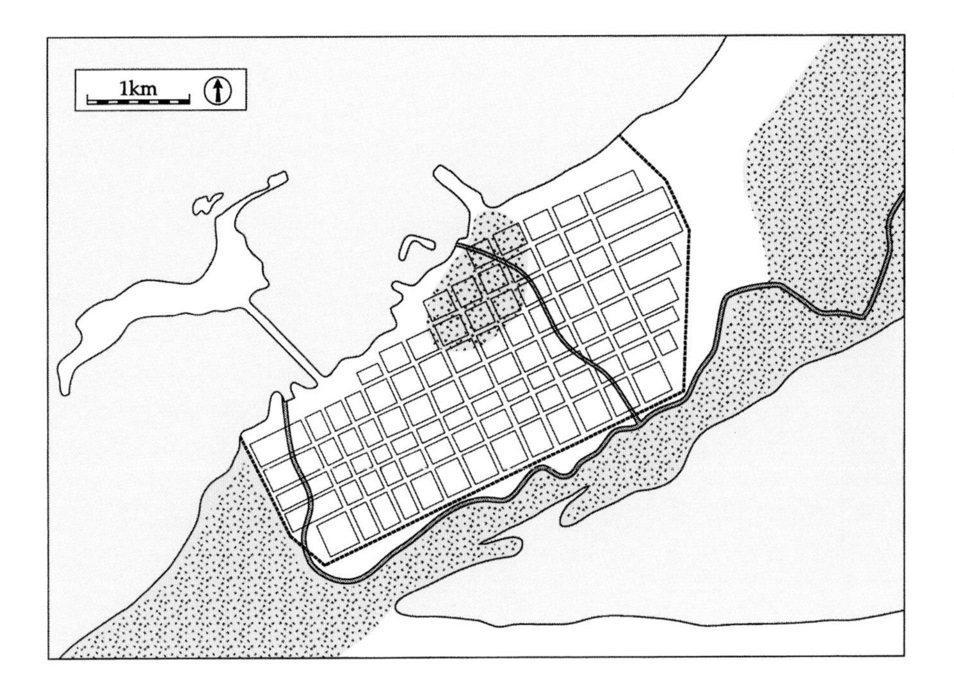

19 Hellenistic Alexandria with its gardens (dotted), both in the urban palace quarter and outside the city wall, near Lake Mareotis and along the Canopus canal (grey).

Egyptian feature harking back to the ancient pharaonic custom of surrounding the palace by gardens, rather than a new feature introduced to Egypt by the Ptolemies.

There are, as yet, no proven Hellenistic forerunners for the pleasure gardens of private Roman houses and villas filled with flowers, trees, shrubs, fountains and statuary. One glance at the houses and towns of the Roman Empire makes it immediately clear how very different their attitude towards nature was. Gardens, whether practical or ornamental, were an integral part of the Roman private house in the city and in the country. Early Roman houses in the fourth and third centuries BC frequently had a small vegetable garden at the rear of the property as testified by excavations at Cosa on the west coast of Italy. Kitchen gardens existed behind all the houses erected in the town in the third century BC.[16]

By the second and first centuries BC, the peristyle courtyard, a Greek architectural form, became a popular feature of aristocratic Roman houses.

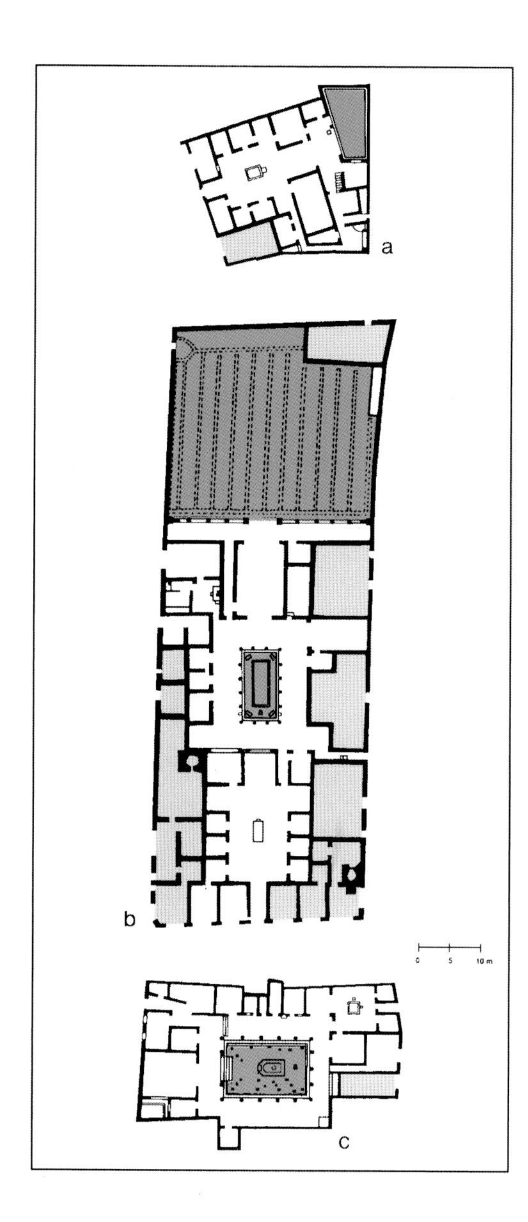

20 Roman houses in Pompeii with areas given over to gardens (green): a. House of the Surgeon, b. House of the Pansa, c. House of the Golden Cupids.

Unlike the Greeks, however, the Romans converted this open area into a garden (fig. 20). At the same time, the small kitchen garden diminished in importance and was, in some cases, abandoned. A number of second-century houses in Pompeii, however, continued to have a utilitarian garden in a walled area at the back of the house, despite also having a peristyle courtyard in the centre of the house. The House of the Faun and the House of the Pansa, for example, appear to have had an orchard or perhaps a mixed garden of fruit trees and vegetable beds at the very back of the house, while the peristyle courtyard in the middle of the house probably had a rather more ornamental character (fig. 21). On farmsteads in the countryside, kitchen gardens were situated for easy access, as we can see at Boscoreale, outside Pompeii, where the owner of the farm laid out vegetable beds next to the main entrance of the house, while he planted the rest of the property with vines.[17]

Gardens in the peristyle courtyard of a Roman house were not necessarily formal or ornamental. The House of Polybius at Pompeii, for example, had a peristyle garden which, as excavations of the subsoil and the root cavities have shown, was an informal, utilitarian garden with fig, cherry, pear and apple trees in it.[18] If the courtyard garden was formally planned and designed to be decorative, it indicated that the owner of the house was wealthy and socially elevated. Those who emulated the aristocratic lifestyle would strive to impress with an ornamental garden, as is apparent in the House of the Vettii in Pompeii (fig. 22). The *nouveau riche* Vettii brothers who acquired and decorated the house in the middle of the first century AD literally covered the walls of the interior rooms with expensive, but not always tasteful, paintings, and they devoted

21 Central pool surrounded by plantings in the courtyard of the House of the Pansa at Pompeii, first century AD.

22 Peristyle garden in the House of the Vettii in Pompeii, replanted according to the preserved design of the first century AD.

23 Planter boxes around the central impluvium *in the atrium of the House with the Relief of Telephos in Herculaneum, first century* AD.

approximately one third of the entire building to a stunning garden full of ornamental plantings, fountains and statues.

The traditional heart of the Roman house was the atrium, just inside the entrance to the house. Before the construction of aqueducts in the late first century BC in the Vesuvian area, the cistern under the floor of the atrium was used to collect rain water. In some cases in Pompeii and Herculaneum, the atrium was then transformed into a decorative garden, with plant containers or planting boxes arranged around the edge of the *impluvium*, or pool, in the centre of the atrium (fig. 23). The same arrangement can be found around pools in the courtyards of Roman houses in Tunisia.[19] The supply of water to Pompeii also meant that gardens could be irrigated with ease, as witnessed by the lead pipes carrying water to the fountains, pools and plants in the courtyard garden in the House of the Vettii. Pliny specifically mentions that one of the advantages of aqueducts is the constant supply of water not only for public buildings and baths, but also for houses, villas and gardens (*Natural History* 36.24.123).

At Pompeii, gardens attached to houses occupied 5.4 per cent of the excavated area of the city, and large food-producing gardens another 9.7 per cent. Wilhelmina Jashemski, the excavator of most of the gardens in Pompeii, has calculated that over one third of the excavated city was open space. These spaces may have been 'open' but they were certainly not 'empty' (see pp. 14–15); they were flourishing gardens of many kinds as archaeological investigations have shown. Perhaps some of the most interesting gardens in Pompeii are the several large gardens in the south-eastern part of the city which were planted as vineyards, fruit orchards and commercial flower gardens.[20] In one of the large vineyards, 2,014 vines were planted and supported on wooden stakes and both the vine stocks and the stakes left cavities in the subsoil. Even though the aqueduct did not reach this part of Pompeii, the gardens were irrigated with rain water from containers and cisterns requiring a great deal of human effort.

Pliny the Younger's description of his Laurentian estate gives us some idea of a Roman villa garden in which the ornamental and the utilitarian were combined:

> All around the drive runs a hedge of box, or rosemary to fill any gaps . . . Inside the inner ring of the drive is a young and shady vine pergola, where the soil is soft and yielding even to the bare foot. The garden itself is thickly planted with mulberries and figs, trees which the soil bears very well though it is less kind to others.
>
> (*Letters* 2.17.14–15)

His Tuscan villa is no less pleasurable because of its gardens:

> Almost opposite the middle of the colonnade is a suite of rooms set slightly back and round a small court shaded by four plane trees. In the centre a fountain plays in a marble basin, watering the plane trees round it and the ground beneath them with its light spray . . . There is also another bedroom, green and shady from the nearest plane tree, which has walls decorated with marble up to the ceiling and a fresco of birds perched on the branches of trees.
>
> (*Letters* 5.6.20–2)

Pliny's descriptions of his villas are all that survive of them, but archaeology has uncovered physical remains of similarly pleasant garden villas. Perhaps the

24 Oleander and box hedges planted according to the original design at the so-called Villa of Poppaea at Oplontis, first century AD.

clearest archaeological evidence for impressive Roman pleasure gardens can be found at the so-called villa of Poppaea at Oplontis near Pompeii (fig. 24).[21] This type of luxury estate is what Roman authors such as Cicero (*To Atticus* 12.25) and Strabo (*Geography* 5.3.6) referred to as 'properties of pleasure' and 'large and costly residences'. Thirteen gardens in and around the villa have been excavated and the villa has yet to be fully uncovered. Most of the gardens found were formal, including the two large gardens behind and beside the villa. The garden at the rear of the building resembled a park, with central and diagonal pathways bordered by shrubs and trees and with statues interspersed amongst the greenery. Along the east side of the villa was an enormous swimming pool, flanked on the east by a row of thirteen trees consisting of oleanders, lemon trees and large plane trees. In front of each tree was a statue or a herm of gods and heroes. At this site, architecture, gardens and garden adornments, as well as the natural scenery (Vesuvius to the north and the sea to the south), were designed to unite in a spectacular way. This is also illustrated at Livia's villa at Prima Porta outside Rome. Visitors' and residents' eyes would

25 Reconstruction of the small peristyle garden in the Villa of Livia at Prima Porta near Rome, c. AD *40.*

26 Courtyard garden in a pool in the Roman House of the Jets of Water at Conimbriga in Portugal, c. *third century* AD.

have been drawn to the view of the 'tamed' landscape in the small peristyle garden, located close to the southern edge of the villa, with the wild mountainous landscape of the *Mons Albanus* beyond (fig. 25).[22] Like Oplontis, this imperial villa had several gardens, of which only some have been excavated.

The replanting of many of the gardens at Oplontis, replicating the Roman planting pattern and vegetation, has been particularly successful, and it becomes immediately clear when walking along the paths and through the courtyards how intimately related the gardens and their architectural surroundings are. Another evocatively replanted and refurbished Roman pleasure garden, but in an urban context, is that in the so-called 'House of the Jets of Water' at Conimbriga in Portugal (fig. 26).[23] In this house, the ornamental courtyard garden was designed as six masonry planter boxes in geometric shapes in the middle of a pool; 400 lead spouts on the edges of the planter boxes and the pool itself spurted water into the pool.

The garden as an extension of the villa, and the close relationship between the two, is also evident at Roman sites in the European provinces. At the Roman palace at Fishbourne in Sussex, the visitor would have entered the courtyard and been led across the formal garden by a central path to the audience room in the west wing of the palace (fig. 27).[24] The planting trenches, flanking the path across the courtyard, probably once held box hedges and the formal arrangement of the plantings complement the architectural surroundings. It is suggested that the northern and southern peristyle courtyards in the eastern wing of the palace were also designed as gardens and recent evidence recovered from the terrace suggests that this area may have been a 'natural' garden with pools and streams. Small semi-circular basins of Purbeck marble were set into the box hedges and were probably supplied with water by ceramic pipes running beneath the main courtyard.

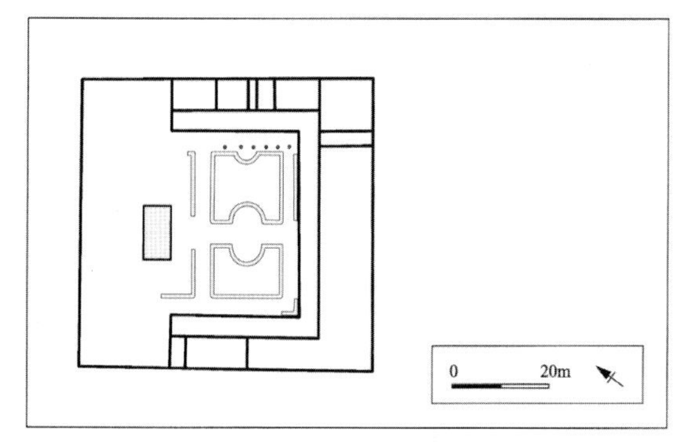

27 (Top) *Model of the Roman palace at Fishbourne, Sussex, with its large formal garden,* c. AD *75.*

28 (Above) *Preserved planting trenches (grey), possibly for box hedges, reveal the design of a garden with a pool next to the Roman villa at Dietikon in Switzerland, first century* AD.

Both the palace and the formal garden are based on purely Mediterranean prototypes, and the entire complex, laid out in this manner in the 70s AD, must have seemed very foreign in Britain and can hardly have failed to impress British visitors who would never have seen anything quite like it on their shores. Whether or not the palace owner was indeed Cogidubnus (or Togidubnus), 'king and legate to the emperor in Britain', who had been granted

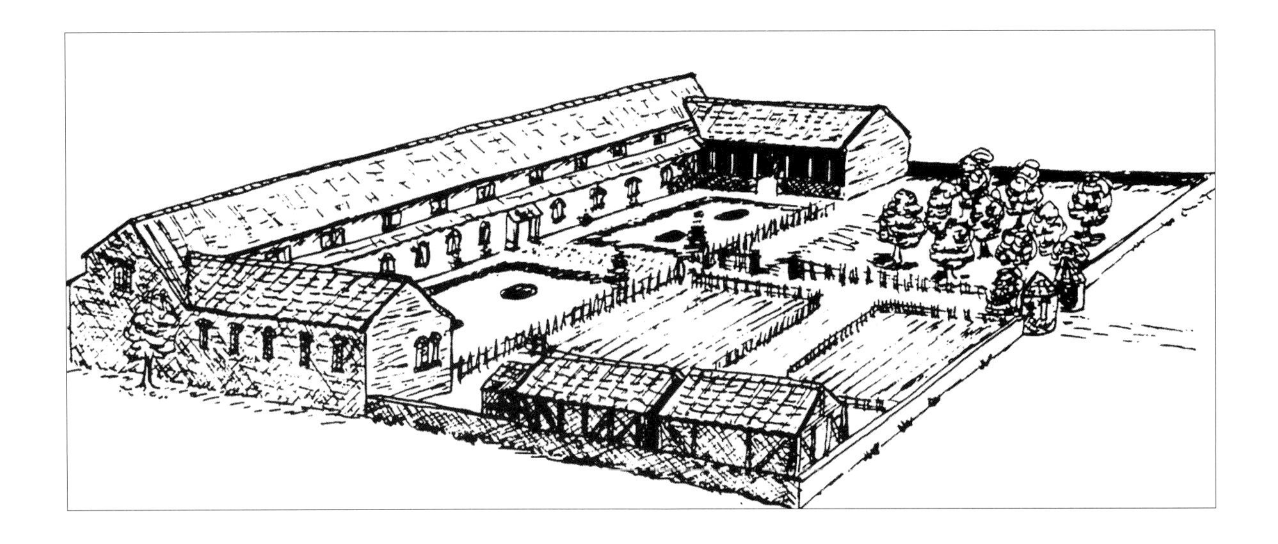

29 Reconstruction of the Roman villa and its garden at Latimer, Buckinghamshire, fourth century AD.

Roman citizenship under the Emperor Claudius, remains a matter of speculation, but the proprietor was certainly a man of wealth and standing in society.

A similar kind of formal and geometric ordering of space with plantings and hedges can be seen at the later first-century Roman villa at Dietikon in Switzerland (fig. 28), and, as at Fishbourne, it is the planting trenches dug into the subsoil, probably for box, that can be used to reconstruct the design of the landscaping in the courtyard.[25] In addition to the formal gardens, which were 'exported' to conquered regions outside Italy in the Roman period, evidence exists to indicate that utilitarian gardens became a common feature of Roman urban houses and farms in the provinces. This evidence includes parallel rows of planting beds for vegetables or small trees in front of the Roman villa at Latimer in Buckinghamshire (fig. 29), deposits of enriched garden soil in a walled enclosure outside the back door of a Roman farm house at Winden am See in Austria, and rows of soil mounds in market gardens outside the city wall at Roman Colchester in Essex.[26]

CHAPTER THREE

Orchards, groves and parks

In the hot climate of ancient Egypt, trees were particularly important for both shade and fruit.[1] Sycomore fig trees were personified as goddesses in Egyptian art, and were perceived as dispensers of nourishment. The tree goddess is often shown in paintings and reliefs as a small female figure crouched in the top of a tree or as a standing figure from whose head grows a small sycomore tree (fig. 30). In her outstretched hand, the tree goddess may hold vegetables, bunches of grapes, baskets of fruit and flowers bestowing them on the owner of the garden. A petition recorded in an Egyptian tomb expresses the wish that 'I may walk each day continuously on the banks of my water, that my soul may repose on the branches of the trees that I have planted, that I may refresh myself under the shade of my sycomore'.[2] The popularity of sycomore figs with humans appears to have been shared by animals such as hamadryas baboons who, in Egyptian tomb paintings, are occasionally depicted perched in a sycomore tree feeding on the fruits (fig. 31).

Other fruit-bearing trees depicted frequently in Egyptian art are the pomegranate and the date palm; these are often arranged around a pool in the garden or orchard near the house. Although no one knows the types of tree that grew in the grounds of the palatial fortress of the Hyksos (in the region of Tell el-Dabaa at 'Ezbet Helmi in Egypt) in 1600–1530 BC, the parallel rows of excavated tree pits immediately to the east of the palace indicate that an orchard or grove formed part of this royal complex.

30 The Egyptian tree goddess gives nourishment to the priestess Henutmehyt on a painted shabti-box, c. 1290 BC.

In the temple of Amun-Re at Karnak, built by Thutmosis III around 1450 BC, reliefs adorning the bottom section of the wall depict a wide variety of trees, not all of which can be identified. These reliefs are commonly known as the 'botanical garden' and the scenes may be based on the observation of

31 Baboons feed on figs in a sycomore tree on a painting in the tomb of Khnumhotep at Beni Hasan, c. *1950* BC.

vegetation witnessed during the Syro-Palestinian campaigns of Thutmosis III.[3] A few years before these campaigns, Queen Hatshepsut had sent an expedition to the land of Punt (upland Ethiopia) to exchange Egyptian commodities for incense and myrrh. Hatshepsut imported not only the resin from myrrh trees, but also had whole myrrh trees, complete with their roots, packed into baskets and brought back to Egypt, almost certainly for the palace gardens or for her monumental funerary temple. The transport of these trees is represented on the reliefs of this temple at Deir el-Bahari (fig. 32). About three hundred years later,

32 Workers carry myrrh trees in baskets from the land of Punt on a relief in the mortuary temple of Hatshepsut at Deir el-Bahari, c. *1470* BC.

the Pharaoh Ramesses III imported incense and myrrh trees and had them planted in Memphis.[4]

The kings of the Near East also imported trees for their royal gardens. Tiglath Pileser I made the following claim around 1000 BC: 'I brought cedars, boxwood and oak trees from the countries which I have subdued, the likes of which none of the kings my forefathers ever planted, and I planted them in the gardens of my land. I took rare garden fruits not found in my own land and caused them to flourish in the gardens of Assyria.'[5] Tiglath Pileser's lead was followed throughout the ninth and eighth centuries BC by Assyrian kings who were known for their interest in gardens. Both Ashurnasirpal II and Sargon II, for example, planted parks and groves near Assyrian cities. Ashurnasirpal

33 Flowers and vines are planted between palm and pine trees in a royal garden inhabited by tame lions on an Assyrian relief from the palace of Ashurbanipal, c. 645 BC.

(884–859 BC) constructed a royal garden in Kalach with plants that 'vied with each other in fragrance' and which he called 'the garden of happiness'.[6] In this garden grew cedar, cypress, fir, oak, willow, ebony, terebinth, juniper, nut trees, olive, apple, pear, fig, pomegranate trees and grape vines. Sargon (722–705 BC) established a city, named Dur-Sharrukin (Khorsabad), in a gigantic park resembling the mountains of Amanus: 'I had all the spices of the land of the Hittites and all the vegetation from their mountains planted close to each other.'[7] The local population were taxed in the form of saplings of apple, medlar, almond, quince and plum trees for this park.

The Assyrian garden was essentially a tree garden planted with palms, fig trees and conifers, although it is possible that other kinds of plants, such as vegetables, flowers and small fruit trees, were also cultivated between the rows of trees (fig. 33). The most common form of utilitarian garden in Mesopotamia, however, appears to have been the palm grove. The date palm was to Mesopotamia what the sycomore was to Egypt, and even centuries later, the Greek geographer Strabo admired the palm trees of Mesopotamia, praising their usefulness (*Geography* 16.1.14). According to him, the palm supplied not only dates but also wine, vinegar, honey, flour, fibres and kernels for fuel or animal feed. Evidence in Egypt indicates that the palm, in particular the dom palm, provided the favoured material (leaves) for mats and baskets (fig. 34).[8] Palm groves encircling Assyrian towns were irrigated by canals which led water from nearby rivers to the trees, and both the irrigation and artificial pollination of the trees was the responsibility of a specialist gardener, a *nukarribu*. This procedure involved the transferral of pollen from the flowers of the male trees to the flowers of the female trees between January and March.

The most important tree in Mesopotamia was valued not only for its products, but also for the cooling, green environment it created when planted in groves. It is this verdant environment on the Tigris and Euphrates rivers that the Book of Genesis claims for the garden of Eden, and it is the fruitful orchard or grove that the biblical Adam and Eve inhabit. 'And out of the ground made the Lord God to grow every tree that is pleasant to the sight, and good for food; the tree of life also in the midst of the garden . . . ' (Genesis 2.9). In the Eastern Church of Byzantium, particularly in fifth- and sixth-century mosaics in Jordanian churches, the garden of Eden is frequently represented as a garden of fruit trees in which the four rivers of paradise flow (fig. 35).

The word 'paradise' comes from the Greek word *paradeisos*, a name the Classical Greeks used to describe the royal parks of the Persian kings.[9] In turn, *paradeisos* is derived from the Persian *pairidaeza*. From the sixth century BC, these royal parks were planted with trees and inhabited by animals hunted by the Persian kings. Astyages gave his grandson Cyrus the Great 'all the game present in the *paradeisos*' at this time so that he could learn to hunt (Xenophon, *Cyropaedia* 1.3.14, 8.1.34–8). This tradition continued well into the seventh century AD when the Sassanian Dynasty in Persia laid out vast parks replenished with elephants, wild boar, deer, birds and water fowl.[10] The historian Xenophon, who accompanied Greek mercenaries employed by the Persian King Cyrus the Younger of the Achaemenid dynasty at the end of the fifth century BC, recorded various stories and tales of Persian royalty. In these accounts, the

34 (Overleaf)
Dom palms, date palms and sycomore trees on an Egyptian painting in the tomb of Sennedjem in Thebes, c. *1279* BC. *The dom palms have split trunks.*

many straight rows of trees in the royal park of Cyrus in Sardis, as well as the large and beautiful park of Belesys, the Achaemenid satrap of Syria, are praised. Cyrus created *paradeisoi* which were filled with 'all the good and beautiful things that the earth wishes to put forth' and in which he spent 'most of his time except when the season of the year prevents it' (Xenophon, *Oekonomikos* 4.13). Belesys' garden was 'a very large and beautiful *paradeisos* with all the products of the seasons' (Xenophon, *Anabasis* 1.4.10).

35 Paradise on a floor mosaic in the Chapel of the Martyr Theodore at Madaba in Jordan, AD 562.

Cyrus the Great (559–530 BC) established a new royal city at Pasargadae in south-western Iran in his lifetime, and in this city was an extensive royal *paradeisos* surrounding the palaces. Excavations in the 1960s helped clarify how the formal gardens at the palaces were laid out.[11] A palace overlooked the vast principal garden (230 × 200 metres) which was surrounded on three sides by broad paths and stone watercourses (fig. 36). Access to this inner garden was indirect: the visitor was led through a gate, across a bridge and through a garden pavilion before catching a glimpse of the garden and approaching the king's throne in the portico of the palace. The Persian royal garden throughout its history made use of pools and axial canals, and especially pavilions, so that the garden could be enjoyed from a sheltered location. The garden at Pasargadae appears to have been divided into four equal plots by a watercourse and a path, the four parts possibly symbolizing the power of the king as 'King of the Four Quarters'. This classic plan of the quadripartite, enclosed garden later became known as a *chahar bagh*, and was a common design for Islamic gardens from Spain to Mughal India (fig. 37). In these contexts, the concept of the four rivers flowing in paradise may have contributed further to the design of the earthly garden crossed by canals.

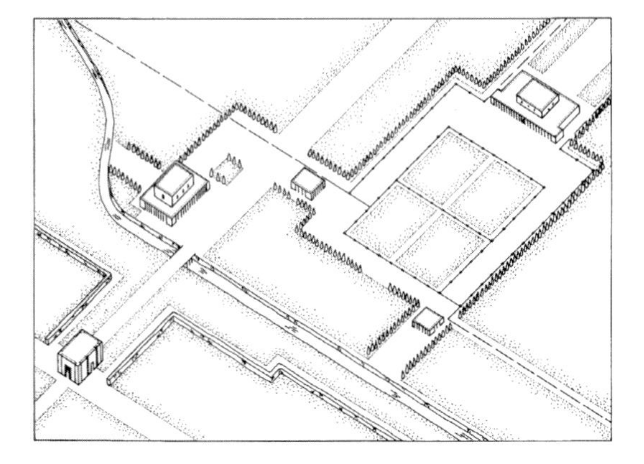

36 Reconstruction of the quadripartite palace garden of the Persian King Cyrus at Pasargadae, c. 550 BC.

After the conquest of Persia by Alexander the Great, the royal parks of the Persian kings were adopted by Alexander, who returned to Babylon before his death in 323 BC. He is said to have bathed and rested in the royal park in Babylon shortly before his demise (Arrian, *Anabasis* 7.25.3). Historical sources record that Harpalos, Alexander's treasurer, introduced Greek plants into these royal gardens at Babylon but that he had no success with ivy as this did not flourish in the hot, oppressive climate of the Persian Gulf (Plutarch, *Alexander* 35.15). After Alexander's death, his successors divided up the Near East and the Greek world, establishing Greek kingdoms in Macedonia, Asia Minor, Syria and Egypt. According to literary sources, one of these Greek kings, Seleukos of Syria, took over the royal *paradeisoi* in Sardis (Plutarch, *Demetrios* 50.6).

These royal gardens and game parks were of great appeal to the Roman upper classes, and wall paintings in Roman houses frequently depicted them. Allusions to the royal park of a Persian or Hellenistic Greek king within the confines of a more modest Roman home formed an ideological link between the owner of the house and the powerful rulers of the Near East who lived in absolute luxury. In fact, the fabulously wealthy Lucullus, a Roman general who fought extended eastern campaigns in the early first century BC, resided in his luxury villa and gardens at the seaside near Naples in the manner of a Persian king, like 'Xerxes in a toga' (Plutarch, *Lucullus* 39.3). Lucullus also established a preserve for wild animals on another of his estates, and Q. Hortensius is said to have dined with his guests in the animal park of his villa near Ostia – the inspiration for which was almost certainly the Persian royal game park (Pliny, *Natural History* 8.78.211, Varro, *Rerum rusticarum* 3.13.2–3).

Royal parks such as these were completely unknown to the Greeks until they encountered them in Persia. In the democratic society of Greece in the fifth and fourth centuries BC, there were no kings and there were no royal gardens. It was not until after the death of Alexander that dynastic monarchies were established by his successors in Greece and the eastern Mediterranean. Thus parks in Classical Greek cities were of a completely different character to those in Persia. The most famous parks of Greek and Roman antiquity were those in which the three oldest educational institutions, known as 'gymnasia', outside Athens were located. The gymnasia in the suburbs – Academy, Lykeion and Kynosarges – were situated in well-watered areas along the Kephissos, Eridanos and Ilissos rivers; all three were established amid ancient sacred groves and cult sites (see fig. 18).[12] The plane, elm, poplar and olive trees in the Academy, in particular, were praised in the ancient sources: 'All fragrant with woodbine and peaceful content and the leaf which the lime-blossoms fling. When the plane whispers love to the elm in the grove in the beautiful season of

37 The quadripartite Islamic garden in the Patio de la Acequia in the Generalife palace in Granada, Spain, divided by a watercourse and a path crossing it (hidden behind the flowering plants), c. AD 1250. The original garden beds lie deep below the soil of the modern replanted garden.

spring' (Aristophanes, *Clouds* 1004–8). Sports grounds, a part of any gymnasium, were often located near wooded areas and water since the athletes and participants required grounds that were shady and cool and also water for the baths. Athletic competitions were always viewed as religious events.

In the fourth century BC, a number of philosophers, such as Plato, Aristotle, Theophrastos and Epicurus, founded their own schools in these wooded areas near the gymnasia. These school gardens were laid out next to the houses of the philosophers, and they were modelled on the larger gymnasium parks on which they immediately bordered. Here again, the garden is to be seen in a religious context. The property of Theophrastos included houses, a colonnade and a garden of the Muses (a *Mouseion*), and the philosopher's own funerary cult was maintained in the grounds. Both Theophrastos and Epicurus bequeathed their gardens to their friends so that they could continue to study literature and philosophy there.

Some of the villas of aristocratic Romans were, in turn, based on the philosophers' gardens. The choice of statuary decoration in these Roman villa gardens, as well as the dialogues and letters of Roman scholars, clearly indicate that many Roman pleasure gardens were based on the Greek gardens in sanctuaries, gymnasia and the schools of Athenian philosophers of the fourth century BC. Cicero named his first century villa near Tusculum his 'Academy', and furnished it with statues, herms and other Greek works of art – including a statue of Plato – which he considered appropriate for a gymnasium (*To Atticus* 1.4, 1.6, 1.8–11). References to the Academy and the gardens of Athenian philosophers are frequent in Roman literature and Roman tourists might have visited these when conducting their grand tour of Greece (Cicero, *De Finibus* 5.1.3). The gymnasia, the *Mouseion*, and art galleries in the groves of the Ptolemaic kings in Alexandria also may have been a source of inspiration for Roman gardens. The wealthy, educated and cultivated Roman discussed philosophy, literature and art with his equally cultivated friends and such discussions frequently took place in their villa gardens.[13] The statuary decoration of the grand Villa dei Papiri in Herculaneum, reconstructed at the Getty Museum in Malibu, indicates that the villa owner carefully selected a variety of copies of Greek works of art thematically linked to gymnasia, libraries and sanctuaries for his peristyle garden (fig. 38).[14] The selection of copies of Greek statues and

herms of gods, heroes and athletic figures along the edge of the swimming pool in the garden at Oplontis also suggests that this assemblage alluded to the gymnasia of ancient Greece.

A closer examination of one Roman urban house, that of Loreius Tiburtinus at Pompeii, illustrates how Roman-villa decoration was adopted on a smaller scale. On the terrace overlooking the garden is a water channel or *euripus* (the Greek word for river or channel, and in this context the Nile river) flanked by statuettes of the Muses, a reclining river god and a sphinx, all

38 Reconstructed Roman garden at the J. Paul Getty Museum in Malibu, California, based on the Villa dei Papiri in Herculaneum.

39 Roman garden divided by a water channel in the House of Loreius Tiburtinus in Pompeii, first century AD.

40 Reconstruction of the forum planted with trees at Cosa in Italy, c. AD 150.

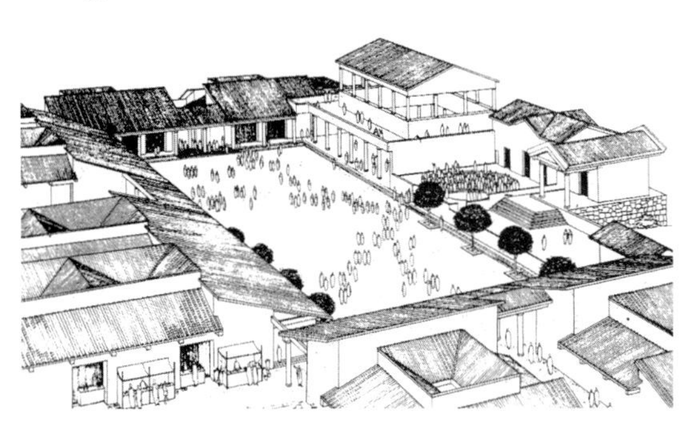

suitable themes for a *Mouseion*, particularly the famous *Mouseion* in Alexandria on the Nile. Other statuettes include a herm, alluding to the sculptural ensembles of Greek sanctuaries and public places, and two groups of fighting animals harking back to Persian royal game parks (*paradeisoi*). The entire length of the garden is divided in two by another *euripus* with fountains, pergolas and statuettes (fig. 39). Thus, in one single garden, references to various types of gardens from various periods in Greece, Egypt and Persia are eclectically drawn together, and these references would have been understood immediately by anyone visiting the garden. Even in the quite modest Pompeian houses of the first century AD, these elements frequently appear either individually or in a combination of motifs, making an overt ideological link to the grand architectural complexes of the eastern Mediterranean and copying the status-linked adornment of the villa estates of wealthy and influential Romans.

In the centre of Greek cities, there was generally not enough space for areas of any size to be planted with trees, although a notable exception is the Hellenistic city of Alexandria where parks surrounded the royal palace. Nevertheless, Greek cities were not completely devoid of trees, since the central square or agora, the heart of political, religious and economic life, could be planted with them. The Athenian agora, for example, was planted with plane trees in the fifth century BC by the statesman Kimon to provide a cool, shady zone for citizens to gather (Plutarch, *Kimon* 13). These appear to have grown on the edges of the square where water from underground drains probably seeped into the ground and allowed vegetation to grow. This corresponds with Plato's suggestion that the overflow or excess water from public fountain houses be piped into the groves and gardens in public places and in sanctuaries (*Laws* 761C). The practice of planting trees in the main public square can also be witnessed at the Roman colony at Cosa in Italy where the earliest colonists in the third century BC saw to the provision of a shady environment by planting trees in pits cut into the underlying rock at one end of the square (fig. 40).[15]

As far as tree plantings in Hellenistic Egypt are concerned, the lines are somewhat blurred between orchards and large plantations based on monoculture. Some of these large plantations are referred to in contemporary papyrus documents as *paradeisoi*, but these are not to be confused with the royal parks or pleasure gardens of the Persian kings. The large estates of Apollonios, the chief financial adviser of King Ptolemy Philadelphos in the third century BC, covered an area of up to 2,756 hectares, and were planted with vineyards, gardens and *paradeisoi*.[16] Those areas referred to specifically as gardens were planted with olive trees, fruit trees and conifers. The *paradeisoi* were quite large and could consist of plantations of up to three thousand olive trees, as well as other cash crops such as roses for garlands and wreaths (fig. 41). The vineyards on these Egyptian estates were not devoted solely to the cultivation of the vine; the papyri reveal that planting beds for onions, melons and garlic were often laid out between the rows of vines.

While orchards and large commercial gardens with fruit trees and vines in Roman Pompeii, as well as the pleasure gardens and parks of private Roman villas, have been discussed in the previous chapter, this chapter looks at the public parks of Rome and the imperial parks. From the middle of the first century BC, wealthy Roman aristocrats vied to outdo each other in ostentatious public display,

41 The deceased on a gilded mummy mask from Egypt holds a garland of red roses in his left hand, second century AD.

particularly in Rome. One of the forms this took was to construct various buildings and surround them with large public parks and squares planted with groves. First to do this was Pompey the Great, a wealthy aristocrat and successful general, who had a grove, theatre, basilica and markets built in 55 BC in the centre of the city.[17] The colonnades ('shady columns') and groves ('avenues thickly planted with plane trees rising in trim rows') built by him were known as the *Porticus Pompeiana* (Propertius 2.32.11–16). Their physical remains lie buried beneath medieval buildings, but the layout of the complex is preserved on a marble plan of Rome prepared during the reign of Septimius Severus in the early third century AD. Adjacent to the theatre was a colonnaded square with a double grove, which drew the eye along the central axis to the theatre (fig. 42).

Pompey, like other Roman generals campaigning in the eastern Mediterranean and the Asiatic provinces, had encountered the parks, palaces and public buildings in both oriental and Hellenistic Greek cities. According to Pliny not only booty and slaves were brought back to Rome during these campaigns but also trees 'figured among the captives in our triumphal processions' (*Natural History* 12.54.11). Pliny specifically mentions trees such as the cherry, the peach, the apricot and the pistachio, which were brought back from the eastern Mediterranean and Syria by aristocratic generals returning to Italy. Sextus Papinius, for example, grew slips of the tuber apple in his camp before returning to Italy in AD 23 (Pliny, *Natural History* 15.14.47). In Pompey's park stood originals or copies of sculptural groups from the eastern Greek world, suggesting that Pompey was linking himself ideologically and politically to the parks and palaces of the Hellenistic kings.

Later aristocrats and emperors continued the tradition of constructing formal public parks in Rome. Vespasian, for example, built his own temple and forum complex, the *Templum Pacis* (later called the *Forum Pacis*), in AD 75 to celebrate his victory over Judea. Recent excavations in the colonnaded square have revealed rows of masonry features, possibly water basins and bases for sculptures, between which were found double rows of planting pots. These contained material that was analysed and found to be the remains of rose bushes.[18] Such public parks and gardens were an integral and important part of the design of the architectural complexes in the very heart of Rome.

42 Reconstruction of the Porticus Pompeiana *planted with trees in Rome, 55 BC.*

43 Aerial view of Hadrian's villa at Tivoli in Italy with its many buildings and gardens, c. AD *118–34.*

In addition to these public parks, a number of wealthy and influential Romans built private villas and parks in or just outside Rome. Agrippa, son-in-law of Augustus, for example, built a villa on the Campus Martius, which included a park, baths and water features. This complex, the *Horti Agrippae*, was left in his will in 12 BC to the people of Rome for their enjoyment, but the owners of many other *horti*, such as Valerius Asiaticus, Maecenas and Sallust, were forced to leave their property as legacies to the emperor.[19]

Imperial interest in huge parks and estates in Rome is exemplified by Nero who took advantage of the fire that ravaged Rome in AD 64 to annex eighty hectares of prime urban land on the slopes of the Palatine and Esquiline hills to construct a new palace, the *Domus Aurea*. This was set amongst a landscaped park of fields, vineyards, pasture, woodlands and a lake. In doing so, he created an artificial rural estate in the heart of the city. The propertied classes of Rome took a rather dim view of this extravagant display of luxury, and when Nero was murdered in AD 68, his successors Vespasian and Titus seized the estate and converted much of it into public property. Vespasian's youngest son, Domitian, rebuilt and aggrandized the imperial palace on the Palatine hill between AD 81 and 96.[20] This so-called *Domus Flavia* included many ornamental gardens and pools, including a hippodrome (race-course) planted as a garden.

The Emperor Hadrian also built a vast imperial residence with parks and gardens between AD 118 and 134, but he chose the countryside as the location for his estate (fig. 43). Built at Tivoli to the north-east of Rome, this estate consisted of a chain of parks with pools, canals, fountains and a wealth of statuary imitating collections of famous Greek artworks. Hadrian was a great admirer of Greek culture and he was well versed in Greek philosophy. His special interest

44 Gardens were planted on either side of the Canopus at Hadrian's villa, modelled on the Canopus canal in Alexandria, c. AD *118–34.*

in Greek philosophy and Athens, a city that was held in high regard by Romans as a centre of learning, is reflected in the inclusion on his estate of parks or groves named 'Academy' and 'Lykeion' after the Athenian gymnasia. The construction of the so-called 'Canopus', a canal and pool with statues such as crocodiles on its banks, in the grounds of the villa, suggests that Hadrian was also linking himself to one of the famous landmarks of Hellenistic Alexandria (fig. 44). These imperial parks and villas were a symbol of wealth, status and power, and they far outstrip the luxury estates of any Roman aristocrat in size, design and opulence.

Whilst these tree-planted areas surrounded the public and private buildings of mortals, the gardens discussed in the next chapter were designed to provide the temples and sanctuaries of the gods with a natural environment.

CHAPTER FOUR

Sacred gardens

The ancient Egyptians believed that the pharaoh was selected by the gods to rule, and with this divine legitimacy the pharaohs were the guarantors of life on earth. The king was so closely associated with the gods that it is difficult to differentiate between the buildings erected to the cult of the gods and those dedicated to the cult of the king. The latter are usually funerary or mortuary monuments, but because of their dual role as temples, these complexes will be discussed here in the context of sacred gardens. The association of vegetation and fertility with the gods and the pharaohs, the givers and preservers of life, is apparent at many of these cult sites.

An axially arranged and paved road or ramp, lined with trees and other plantings, generally led to these monumental mortuary temples. The remains of such groves and gardens in the outer courts of the temples have been found at a number of excavated sites, such as the terraced temples of Mentuhotep II and of Queen Hatshepsut at Deir el-Bahari opposite modern Luxor, both of them in the middle of the desert.[1] It is largely due to this desert location and the aridity of the climate that botanical material has survived. At the temple of Mentuhotep II, built around 2020 BC, fourteen trees were planted in enormous pits up to 10 metres deep at regular intervals on either side of the ramp leading up to the building (fig. 45). The complex, or at least the landscaping of the complex, was not brought to completion, and work appears to have been interrupted during the digging of these planting pits. Trees had been planted only in the pits closest to the temple itself. These pits are now filled with Nile mud, in which the remains of roots, bark and leaves belonging to sycomore and tamarisk trees have been preserved. In the row of trees to the south of the ramp were two rectangular beds of flowers whose stems and roots have been

45 Aerial view of the mortuary temple of Mentuhotep at Deir el-Bahari with preserved planting pits (circular depressions) for a grove of trees in front of the temple, c. 2020 BC.

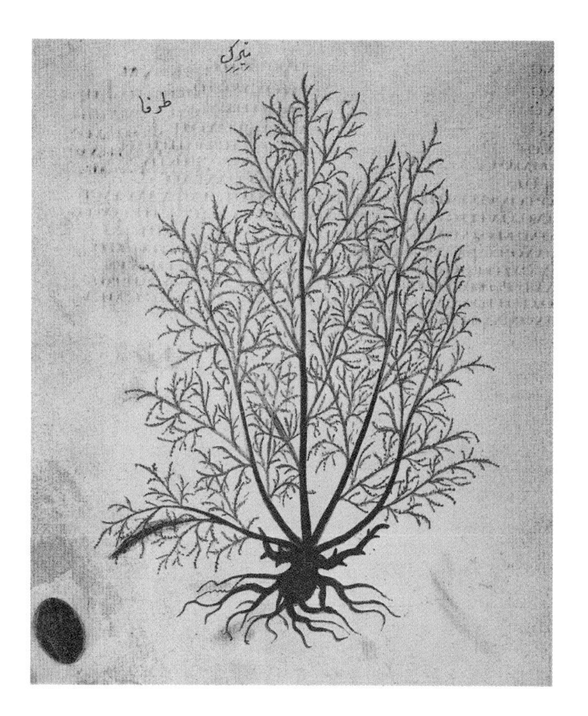

46 Tamarisk in the illustrated herbal of Dioscorides copied in AD *512 in Constantinople.*

preserved. In the area immediately in front of the mortuary temple was a grove of tamarisk trees planted in regular rows in pits up to 2.4 metres in diameter and 1.3 metres in depth (fig. 46). The choice of these two types of trees was deliberate for reasons of religious symbolism. In the sycomore dwelt the sky goddess Hathor, and the tamarisk was the place where the soul of the god Osiris rested and where the pharaoh was reborn as the sun. Mentuhotep, therefore, was associated with the sun which the sky goddess consumed every night and gave birth to every morning.

The temple of Hatshepsut, built around 1470 BC, stood next to that of Mentuhotep II. It was also built as a terraced temple with a garden in the outer forecourt. This garden had two T-shaped pools on either side of the central ramp and, as an analysis of the dried Nile mud deposits on the floor of the pools indicated, they were planted with papyrus. Circular pits were dug around the pools to accommodate trees or bushes.

There is ample evidence for other temple gardens in Egypt. At Amarna, the remains of papyrus and the blue lotus were preserved in the muddy deposits in a building in the Maru-Aten complex that is interpreted as a chapel.[2] At the same site, adjacent to the so-called Main Chapel in the Workmen's Village, was a walled enclosure within which were brick-bordered planting plots containing a layer of dark alluvial soil.[3] There are also wall reliefs in the tomb of Merire at Amarna which show a temple and its garden (fig. 47). In this scene, trees surround the temple and, in a walled enclosure next to the temple, date palms, dom palms, pomegranate trees and vines are arranged around a pool. This arrangement of pool and trees was actually preserved in the funerary temple of Amenhotep, son of Hapu, at Thebes.[4] Amenhotep, who was worshipped posthumously as the god of medicine, was the architect of King Amenhotep III and the 'Leader of the Festival' at Karnak. There were twenty sycomore trees in round planting pits with brick borders in a large courtyard in

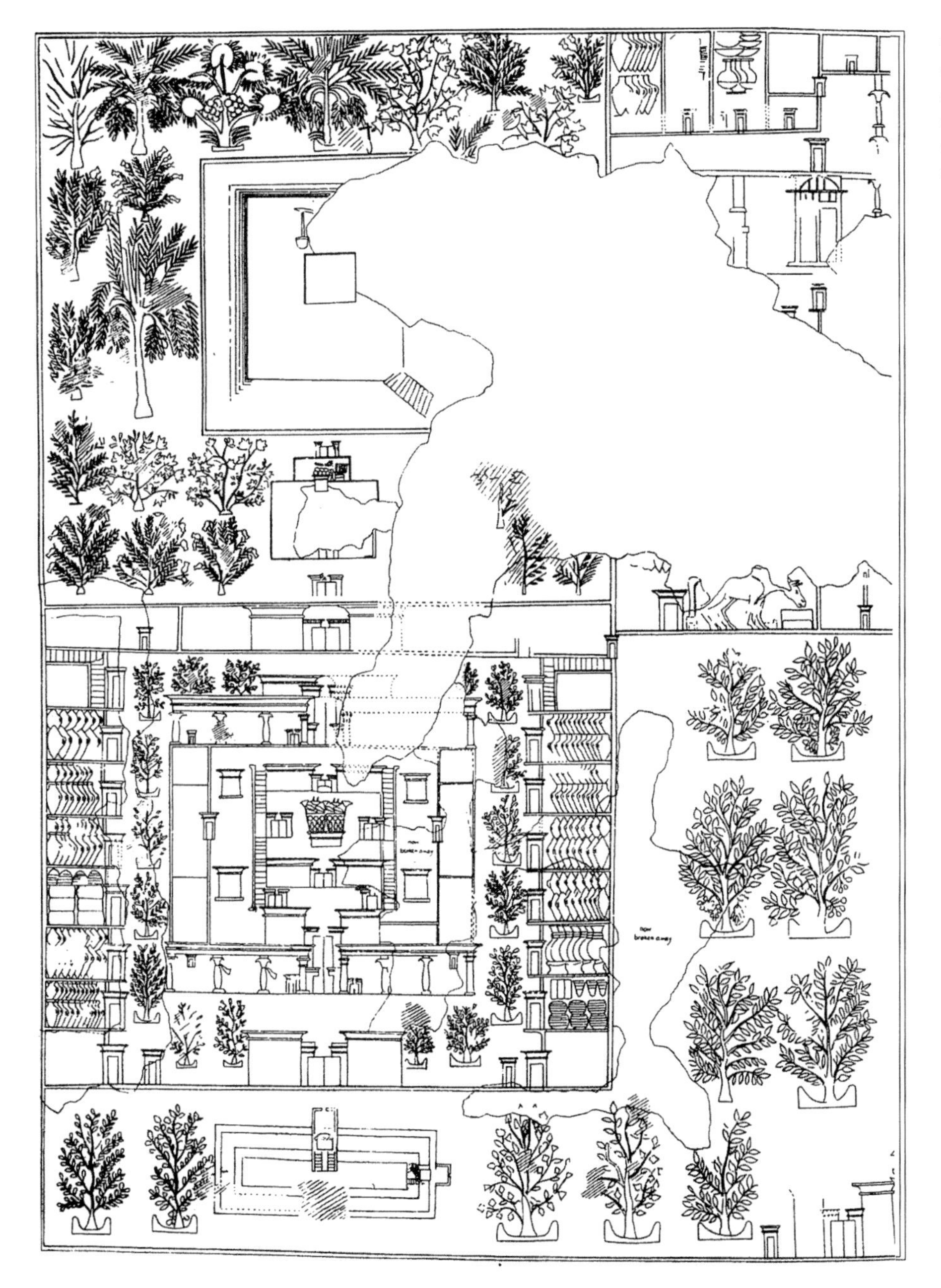

47 An Egyptian temple and temple garden on a relief in the tomb of Merire at Amarna, c. *mid-fourteenth century* BC.

front of his temple, built around 1360 BC. The central pool in this garden had been dug deep enough to reach the water table so that it filled up with ground water.

In addition to their religious connotations, temple gardens in ancient Egypt were also a source of income. Gardens owned by the temples and cults could include vineyards, orchards and groves quite separate from the actual site of the temple. The products from these estates were used in cult ritual and fed the temple personnel, and the income from them supported the upkeep of the cult. A total of 513 gardens and estates donated in the twelfth century BC by Ramesses III to the temples at Thebes, Heliopolis and Memphis are recorded in lists of donations written on papyrus.

Compared to ancient Egypt, the evidence for sacred gardens in the Near East is meagre. Cuneiform texts occasionally refer to the donation of gardens to the gods. Ishme-Dagan, son of Shamshi-Adad, King of Assyria from 1814 to 1782 BC, recorded in a letter to his brother that he intended to plant 'a garden for the god Addu' that should be 'full of juniper trees'.[5] Written sources also record the establishment of sanctuary gardens in the thirteenth and twelfth centuries BC in the Elamite city of Susa in south-west Iran.[6] Only one certain temple garden has been excavated in Assyria, and that is the temple of the god Ashur in the Assyrian capital of Ashur (fig. 48). The building, located to the north-west of the city near the Tigris, had a grove of trees surrounding it and another in the internal courtyard. An inscription carved on an alabaster slab records that King Sennacherib (705–681 BC) had the temple built and that he also saw to the construction of irrigation canals to water the garden and the orchard. Excavations uncovered many planting pits for trees and sections of interconnected canals, and the excavator calculated that originally there might have been more than two thousand trees arranged in rows on either side of the central passage through the courtyard and around the building. The proximity to a source of water was probably an important factor in the choice of location for the temple garden, and perhaps this is the kind of location that we should expect for other Mesopotamian gardens.

God-given were the fruitful trees of Israel, and in the Book of Isaiah (41.19), the Hebrew god promised: 'I will plant in the wilderness the cedar, the shittah tree, and the myrtle and the oil tree; I will set in the desert the fir tree, and the pine and the box tree together.' In another chapter, some of these same trees are said to 'beautify the place of my sanctuary' (Isaiah 60.13), suggesting that gardens were planted in holy places, either around a temple or an open-air altar. Gardens associated with gods other than the Hebrew god are railed against, as are those who 'sacrifice in gardens and burn incense upon altars of bricks' and 'purify themselves in the gardens behind one tree in the midst' (Isaiah 65.3, 66.17). An altar of Baal, surrounded by a grove, is the object of an Israelite attack in Judges 6.23–30 and, according to Josephus (Jewish Antiquities 10.52), the Jewish King Josiah cut down the groves dedicated to foreign gods and razed their altars. Unfortunately, this tells us nothing about the overall appearance or location of such sacred gardens. However, the juxtaposition of altar and trees appears to be a Near Eastern tradition which is also found in the temple (Temple III) at the Canaanite settlement at Tell el-Dabaa in the Nile delta built after 1700 BC.[7] Next to its altar were one or two pits – the acorns preserved in the soil suggest these could have been for oaks.

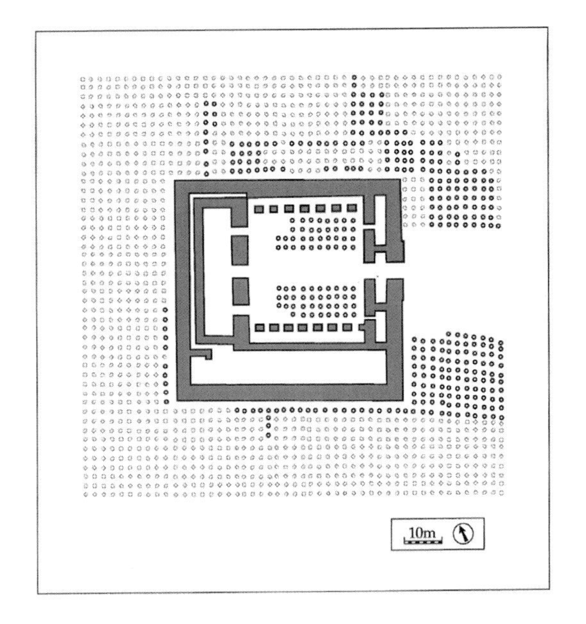

48 A veritable forest of trees in planting pits surrounded the Assyrian temple of Ashur built by King Sennacherib, c. 705–681 BC. The excavated tree pits are shown as black dots, the conjectural pits are grey dots.

Since Prehistoric times, Cyprus has had a lengthy tradition of the veneration of fertility gods associated with gardens, groves and vegetation.[8] The goddess Aphrodite was very important in the religious life of the island and a sanctuary dedicated to her was established around 1200 BC in Old Paphos (Kouklia). This sanctuary is situated approximately 8 kilometres inland from the south coast where, in mythology, the goddess was said to have risen out of the sea onto the land, where she left flowering plants in her footsteps. According to the Greek historian Herodotos (*History* 1.105), this Cypriot cult had originated in Syria where Astarte, the fertility goddess of the Near East, was venerated. Well into the Roman period, pilgrims from all over the ancient world travelled to the sanctuary, particularly during the festival of the goddess, starting their procession from the coast and ending it at Hierokepis, the sacred garden. Just outside Kouklia is the modern village of Yeroskipos, which

49 Excavated water channels and planting pits for trees in a temple grove at Kition in Cyprus, c. 1300 BC.

preserves the ancient name of Hierokepis since Yeroskipos is its modern Greek pronunciation. There is, as yet, no archaeological evidence for gardens or plantings at the sanctuary, so we are dependent on literary references to a sacred garden.

In the ancient city of Kition (modern Larnaca), however, excavations have uncovered the physical remains of a temple garden. After about 1300 BC, two temples were built on either side of the garden with a central pool fed by water from wells and water channels (fig. 49). The modest size of the 116 planting pits suggests that only small trees or bushes could have been planted in this sanctuary. Who was actually worshipped in the temples and the garden is unknown, but at least in the period after 1000 BC, it was the fertility goddess Astarte for whom the Phoenicians erected a third temple on the site.

A third Cypriot cult site with evidence for gardens is the sanctuary of Apollo Hylates, or 'Apollo of the Woods', at Kourion, to which many pilgrims travelled, especially in the Roman period.[9] At some point after the seventh or sixth century BC, trees or bushes were planted in pits and trenches cut into the bedrock along the edges of the sanctuary precinct, but by the first century BC this vegetation must have ceased to exist since the rock-cut pits and trenches were paved over or built upon. In addition to the planting of vegetation on the borders of the precinct, a circular monument was erected that enclosed six pits, possibly for living or artificial trees. This was in use in the Hellenistic and Roman periods, at least until the second century AD. Terracotta figurines of the late seventh and early sixth centuries BC found at Kourion and other Cypriot sites depict three figures dancing in a circle around a tree – these may be associated with a particular ritual involving sacred trees that was carried out at the circular monument in Kourion (fig. 50). The description by Aelian (*Nature of Animals* 11.7) of the grove of Apollo at Kourion in the second century AD as an extensive grove filled with wild animals, cannot be related to the plantings within the sanctuary proper. However, it is likely that the cult owned woodlands separate from the temple site, especially since Apollo was revered at this site as the god of woodlands.

50 Figures dance in a circle around a sacred tree in a terracotta figurine from Chytroi, possibly reflecting a ritual enacted at the temple of Apollo Hylates at Kourion in Cyprus, seventh/sixth century BC.

Groves commonly surrounded ancient Greek temples, and even in the second century AD Roman travellers to Greece could still admire the sacred groves of olives, pines, cypresses, oaks, laurels and fruit trees at many sanctuaries. Certain types of trees were sacred to particular gods: the oak was associated with Zeus, for example, the laurel with Apollo and the myrtle with Aphrodite. The last had a sanctuary in Classical Athens, which was known as the 'Sanctuary of Aphrodite in the Gardens'. She was also connected with flowers and occasionally her son Eros is depicted in Athenian vase paintings as a sacred gardener who waters the flowers in the garden (fig. 51). In the pre-Classical periods, sacred groves were normally natural woodlands, but from the fifth century BC, trees were often intentionally planted around new temples.

51 Eros as a sacred gardener waters plants in an Athenian vase painting, fifth century BC.

The image of the ancient, gnarled tree in the sanctuary, in whose shade worshippers perform a sacrifice, was a popular one in Hellenistic votive reliefs (fig. 52).

In some cases, archaeological evidence corresponds directly to literary references. Pausanias (2.1.7) mentions the cypress grove at the temple of Zeus at Nemea, for example, and excavations have revealed twenty-three planting pits dug into the living rock.[10] In other cases, temple groves are not mentioned in the written sources, but their material remains have survived indicating they did indeed exist. This is the case at the temple of Asklepios in Corinth, where seven planting pits for trees have been found north of the temple.[11] Equally, the reference to a 'grove' of olive and laurel trees surrounding the Altar of the Twelve Gods in the Athenian agora seems slightly exaggerated, given that excavations conducted there revealed the remains of only three or four trees (Statius, *Thebaid* 12.481–496).[12]

As in ancient Egypt, the cult organizations in Classical Greece possessed agriculturally productive estates, which could be located at a considerable distance from the cult building itself. Numerous inscriptions found in excavations on the island of Delos record the estates that belonged to the cult of Apollo.[13] These included gardens, fields and houses, and the inscriptions of the Hellenistic period inform us that the gardens were usually planted with fig trees, olive trees, apple trees, nut trees, myrtles and vines. On the neighbouring island of Rheneia were ten estates, on Delos seven, and in the Hellenistic period further estates were acquired on the island of Mykonos, giving a grand total in the third century BC of twenty-three estates. As is the case with all temple estates, those of Delian Apollo were leased to individuals who worked the estates and ensured that the gardens produced a surplus, the lessee also keeping a part of the produce for himself and often having the right to live on the property.

52 An ancient plane tree offers shade to worshippers in a sanctuary on a Greek votive relief, second century BC.

The adornment of temple precincts with groves was also common practice in Roman Italy and the Roman provinces, although most of the evidence for temple gardens comes from literature. One of the earliest archaeological examples of a temple garden is that at Gabii in Italy, dating to the later second century BC. Excavations at the site, 19 kilometres east of Rome, revealed rows of square holes cut into the rock for planting shrubs and trees between the

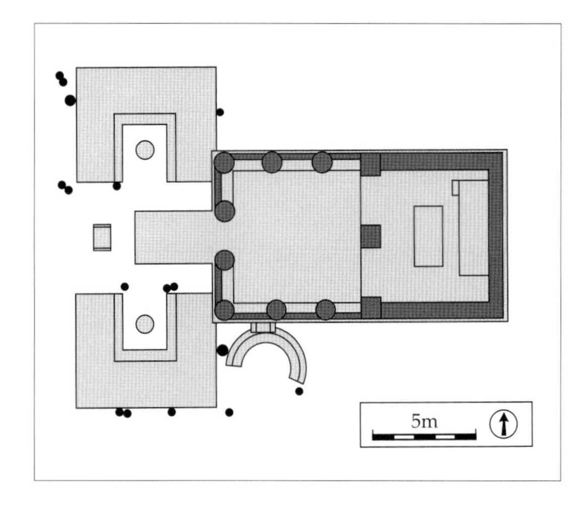

53 (Above) *Vine pergolas shaded the horseshoe-shaped dining couches and the semicircular bench at the temple of Dionysos in Pompeii, first century* AD. *The dots indicate root cavities of vines.*

54 (Below) *The white blossoms of the flowering myrtle were sacred to Venus.*

temple itself and the eastern precinct limit.[14] In Rome, remains of soil and roots embedded in rows of travertine marble containers on the west and north side of the temple of the deified Julius Caesar suggest that these containers held plantings.[15] While there is some evidence of Pompeian temple gardens, most archaeological attention has been focused on private gardens. The temple of Dionysos outside the southern city wall of Pompeii was excavated in 1947–8, but it was not until 1973, when the courtyard of the sanctuary was re-excavated by Wilhelmina Jashemski, that evidence for plantings was retrieved.[16] In the courtyard were masonry dining couches or *triclinia* used for Dionysiac festivals, and the *triclinia* were surrounded by root cavities and post holes that suggest that a vine arbour was planted around the couches to provide the participants with shade (fig. 53). Evidence also suggests that the rear part of the precinct had been planted with rows of vines.

Also at Pompeii, my excavations in 1998 in the courtyard of the temple of Apollo uncovered possible planting pits along the east side of the temple.[17] Preliminary, as yet unpublished, excavations in the same year at the temple of Venus, however, produced much clearer evidence that this precinct had been planted with trees and shrubs, possibly her sacred myrtle bushes (fig. 54), when the temple was built after 80 BC. This temple garden was a manifestation of the fertility guaranteed by Venus. In the prehistoric and historic periods in the Near East and Cyprus, she was venerated either as Astarte or Aphrodite, the goddess of fertility, and her temples were surrounded by gardens. The largest temple in

Pompeii, dedicated to this patron goddess of the city, could hardly have failed to possess gardens.

Finally, Jashemski also excavated the remains of a Roman temple garden at Thuburbo Maius in Tunisia.[18] As at the temple of Venus in Pompeii, root cavities of trees were found in the courtyard precinct between the temple and the surrounding colonnaded halls. The size of the root cavities suggested that trees or large shrubs were planted in them. Literary references to the groves and gardens of the temple of Apollo at Daphne outside Antioch in Syria indicate that temple gardens were highly valued and maintained in the Roman world as late as the fourth and fifth centuries AD. The groves at Daphne are even depicted as a landmark of Antioch on a fourth-century illustrated map of the ancient world, known today in a medieval copy as the *Tabula Peutingeriana* (fig. 55).[19]

The groves and gardens discussed here were an integral part of religious life in the ancient world. They were described in literary sources, depicted in pictorial sources and have been the focus of archaeological investigations. Closely related to the sacred gardens are the tomb gardens connected with the cult of the dead, which are discussed in the next chapter.

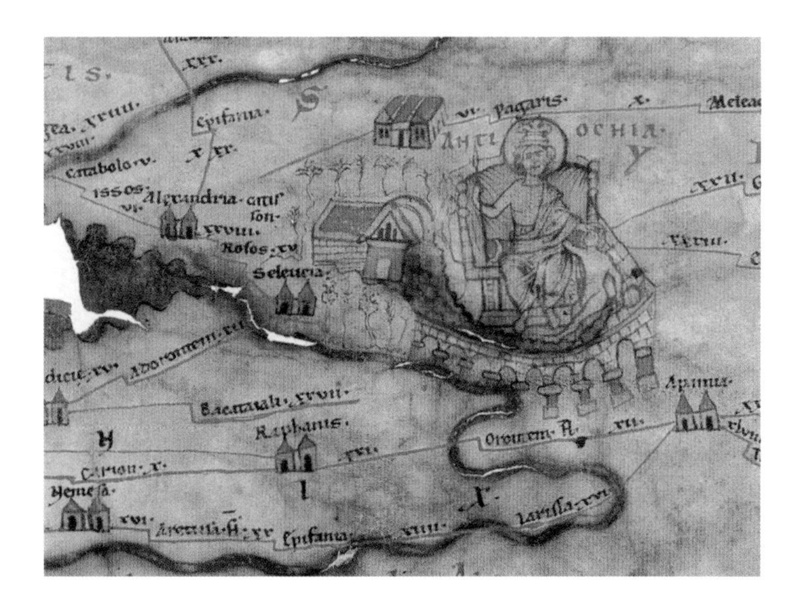

55 Groves and gardens at Daphne in Syria are represented on the Tabula Peutingeriana, *a Medieval copy of a Roman map of the fourth century AD.*

CHAPTER FIVE

Gardens of the dead

The divide between the dead and the living in ancient Egypt was not profound. Inscriptions and paintings in Egyptian tombs clearly show that the deceased hoped to return to earth to be able to enjoy his possessions, and these included, above all, the garden. It was a refreshing and well-watered garden, a garden that could flourish on its own and without any human effort. The mortal Egyptian left this world and entered the other world of eternal happiness, a world characterized as a garden paradise. The eastern horizon of heaven, where the sun rose between sycomore trees, was a symbol of this entry into paradise, and fruit-bearing trees native to Egypt, and at home in any Egyptian garden, gave the deceased nourishment (fig. 56). As the tree of the goddess Hathor, the sycomore protected and fed the deceased in her name. The depiction of an ideal garden, in which the afterlife could be spent, can be interpreted as a kind of 'magical formula' designed to fulfil the wish of the deceased. Wooden models of houses and gardens, like that in the early twelfth-century tomb of Meketre in Thebes, may have had the same function (fig. 14).

Actual gardens at Egyptian tombs would have been difficult and costly to maintain if they were not along the banks of the Nile or in the delta region, where they could easily be fertilized and watered. A painted wooden slab of around 900 BC from western Thebes shows the location of tombs in the western hills in the desert, whilst beyond that, closer to the Nile, gardens with fruit trees flourish (fig. 57). Tombs with gardens, groves and irrigation, particularly in the desert, would have remained a luxury that only the wealthy and powerful could afford. In the last chapter, some of these early monumental mortuary temples in the desert at Deir el-Bahari have been discussed. By the Nineteenth Dynasty, of the early thirteenth century BC, private tombs at Thebes were often

56 An Egyptian garden planted with fruit trees around a pool in a painting in the tomb of Nebamun at Thebes, c. 1400 BC. Note the tree goddess in a sycomore tree in the upper right corner.

57 Fruit trees flourish near tombs on a painted wooden slab from Thebes, c. *900* BC.

58 Oxen pulling a water wheel, or saqiya, *for the irrigation of gardens and orchards in a Roman tomb painting from the Wardian cemetery in Alexandria,* c. *second/third century* AD.

built with a walled outer courtyard in front of the actual tomb.[1] The central feature of this courtyard was often a small bed for plants or possibly a single sycomore tree. This was a garden *en miniature*; it had more of a symbolic value than anything else and it would have been fairly easy to care for.

As late as the Roman period, Egyptian tombs were painted with scenes of gardens, albeit less stylized ones. One of these in the cemetery of Wardian outside Alexandria is painted with a scene of palms and other trees, and with oxen who pull a waterwheel (fig. 58).[2] Gardens planted in the cemeteries of Ptolemaic Alexandria are recorded in Greek papyri of the first century BC as *kepotaphia*, or tomb gardens.[3] Further details, such as the type of plants grown in the tomb gardens, are also recorded and they include fruit trees and vegetables. Moreover, these *kepotaphia* were leased for a certain price and for a certain period of time, so that the profit from the gardens could be invested in the upkeep of the tombs. According to Strabo (*Geography* 17.1.10), there were numerous gardens west of Alexandria and, not surprisingly, they were located near a natural source of water.

Homer's *Iliad* (6.416–420) contains references to the funerary groves of heroes of a bygone era: 'He killed Eetion but did not strip his armour, for his heart respected the dead man, but burned the body in all its elaborate war-gear and piled a grave mound over it, and the nymphs of the mountains, daughters of Zeus of the aegis, planted elm trees about it.' Perhaps such heroic tombs later served in some ways as models for the tombs with gardens and groves in Classical Greece. In his vision of an ideal society, Plato (*Laws* 12.947) imagined the provision of special burial places for those who deserved particular honours, perhaps considered to be heroes in their own time. In this ideal society, according to Plato, such burial places should consist of a grave mound with a grove – both set within an enclosure wall. Annual celebrations with musical and gymnastic competitions and chariot races were to take place in honour of the deceased. In Plato's day funerary celebrations were a regular part of the cult of the dead, even though the law did not permit them to be as extravagant as those that Plato proposed, and there were indeed monumental tombs on plots of land enclosed by walls in the Athenian cemeteries of the fifth century BC (fig. 59). Although it has been assumed that plantings existed within these enclosure walls, there is no physical evidence of them.

59 The Classical tombs in the Kerameikos cemetery at Athens were surrounded by plots that were possibly planted as tomb gardens.

By the Hellenistic period, however, monumental tombs with accompanying gardens within enclosure walls were fairly common in Asia Minor (modern Turkey) and Egypt. Through the construction of monumental tombs and gardens, the dead received cult status and were revered by succeeding generations.[4] A number of documents, recording the last will and testament of such people, have survived: private houses, tombs and gardens were left as legacies. In order to raise funds for the expense of the maintenance of the tomb and the cult, and for the funerary banquets in memory of the deceased, gardens were sometimes leased to individuals for an agreed price. Examples of this practice are recorded in the third and second centuries BC at Halikarnassos and Kos, and even before this in the fourth century BC the Athenian philosopher, Theophrastos, established in his will that his tomb should be erected in his school and garden in the suburbs of Athens, and both the garden and the tomb were to be maintained by his followers.

Monumental tombs in the form of chamber tombs in barrows were common in fourth-century Macedonia, particularly for members of the royal family who were buried in heroic style. A number of these have been excavated in Vergina, ancient Aigai, the capital of the kingdom of Macedonia. Whether these tombs of the royal family were also provided with groves and gardens is unknown, but the painted scene on the façade of a chamber tomb belonging to Philip II of Macedon (359–336 BC) or Philip Arrhidaios (323–317 BC) depicts a hunting party in a grove of trees (fig. 60).[5] The painting probably represents Philip II and his heir and successor, Alexander, as well as court nobles engaged in this royal activity, and it may allude to the royal hunting parks of the Persian kings. According to Arrian (*Anabasis* 4.13.1), it was customary from the time of Philip for young Macedonian noblemen to accompany the king in Persian fashion on the hunt. The setting is not a specific grove or game park; the votive plaque fastened to one of the trees and the pillar surmounted by three statuettes suggest that the hunt is taking place in a sacred grove, which in reality would not have been used for hunting, at least in Greece. It is an idealized landscape, but one that was influenced by the wooded environments in Greece and Persia familiar to the viewer.

The Romans valued gardens at their tombs since they were so important to them in life, and the souls of the dead were almost certainly thought to enjoy them when they 'visited' the tomb. Life did not cease when one died, it simply took on another form: 'Sprinkle my ashes with pure wine and fragrant oil of spikenard; bring balsam, too, stranger, with crimson roses. Unending spring pervades my tearless urn. I have not died, but changed my state' (Ausonius, *Epitaphs* 31). Many epitaphs from Roman tombs mentioning gardens have survived, although unfortunately almost all the tombs to which they were attached have not.[6] The main component parts of such tombs appear to have been the tomb itself, an enclosure wall and a garden, although some of the

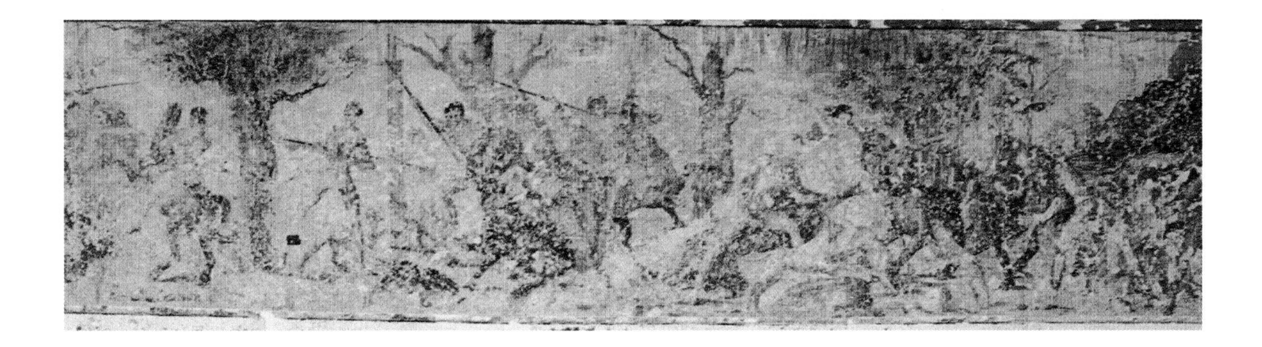

60 Painted frieze on the façade of a tomb at Vergina, Greece, showing a royal hunting party in a grove of trees, c. 336–317 BC.

epitaphs also record buildings for funerary feasts, as well as orchards and vineyards. This is in keeping with the fictional and fabulously wealthy Trimalchio who wished to have 'all kinds of fruit growing around my ashes and plenty of vines' (Petronius, *Satyricon* 71). In some instances, the size of the garden around the tomb is recorded on inscriptions.

As in the Hellenistic Greek East, gardens, orchards, vineyards and estates, as part of the legacy of the deceased, had the practical purpose of providing for the upkeep of the tomb and the financing of the commemorative festivals at the tomb. Such provision for gardens and post-mortem honours are recorded in a funerary inscription of a man who wanted his gardens to provide enough income so that his survivors 'may offer roses to me on my birthday forever; gardens which, I direct, shall not be divided up or misappropriated'.[7]

A rare marble slab from a cemetery on the Via Labicana in Rome actually depicts a stylized plan of a funerary monument and accompanying gardens (fig. 61). The entire complex borders the street, exposing it to as many passers-by as possible. The grieving Cicero sought a prominent location for the tomb and grove of his beloved daughter Tullia, who died in 45 BC, so that it could be visited easily and frequently by those wishing to preserve her memory (*To Atticus* 12.36–7). Visitors to the tomb, who kept the *memoria* of the deceased alive, were an important aspect of the funerary cult. In the painted scenes decorating the walls of the tomb of Patro in Rome, a procession of visitors makes its way past trees of various kinds to his tomb.[8] The main scene shows an ideal garden with fruit trees and songbirds. A poem on the tomb walls reinforced this idealized image of a garden where thorn bushes had no place.

61 Detail of a plan on a marble slab found in a cemetery on the Via Labicana near Rome showing the design of a Roman tomb (circular monument) and its gardens consisting of square and rectangular planting beds and rows of trees (dots). The burial plot borders on two streets (top and right).

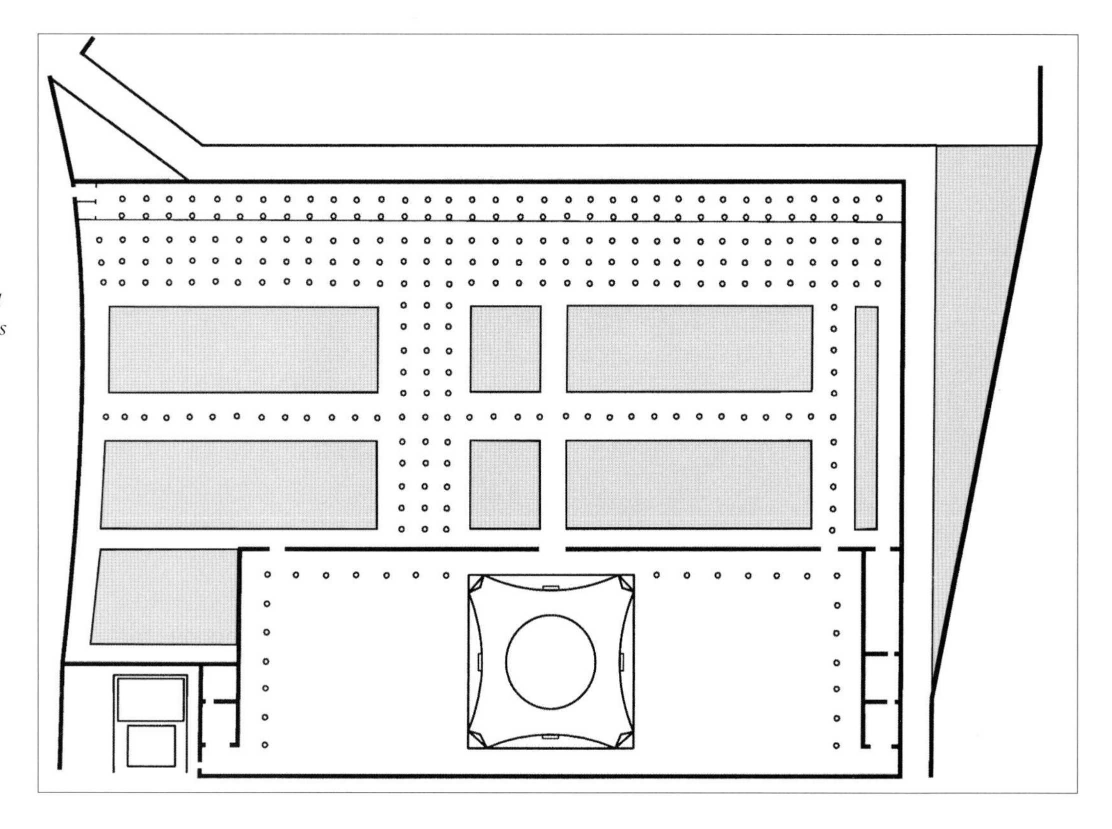

Grand commemorative monuments were beyond the reach of many, but for those who could afford them, these tombs expressed the status of the deceased in perpetuity. This is particularly true for the tombs of the Roman emperors, yet apart from the spectacular funerary buildings, there is little else known about the immediate surroundings of the tombs of the imperial house. The mausoleum of Augustus in the Campus Martius in Rome, built by 27 BC, was an enormous circular building approximately 90 metres in diameter.[9] Various reconstructions of the superstructure of the mausoleum show an earth mound planted with cypress trees on top of the circular socle, but for this there is no concrete evidence. If plantings were associated with the mausoleum, these were the public parks, mentioned by Suetonius (*Augustus* 100.4), through which the visitor gained access to the building.

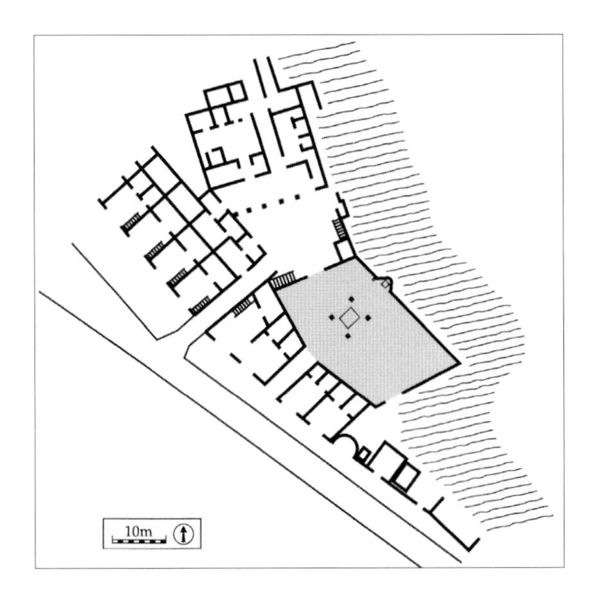

62 Outside the Herculaneum Gate at Pompeii, tombs (lower right), shops (along the road) and a private house (top) were arranged in close proximity to each other and to a garden (grey) with trees, flower-beds and water features, first century AD. *Unexcavated areas are marked with wavy lines.*

It is surprisingly difficult to find physical remains of plantings for tomb gardens anywhere. Again, it is Pompeii, with its very favourable conditions for the preservation of gardens, that offers this form of archaeological evidence. At Pompeii, numerous tombs – often enclosed in a plot of land – flanked the roads leading in and out of the city.[10] One of these tombs outside the Herculaneum Gate was a semicircular structure with a plot of land behind it, where the remains of a tree have been found (fig. 62). Further down the road stood a villa, and between the villa and the tomb was a garden with trees, flower-beds and various water features. The garden could be directly accessed from both the villa and the tomb, indicating the intimate connections between the spaces for the dead and the living.

None of the gardens and groves in the preceding chapters could have flourished without the help of gardeners and agricultural labourers who planted, watered, weeded, trimmed and harvested the plants. In the next chapter, gardeners in the ancient world and their 'tricks of the trade' are discussed.

CHAPTER SIX

Gardeners and gardening

The status of a gardener in ancient Egypt ranged from that of a simple labourer, who worked under the supervision of a foreman, to that of a high-ranking official administrator. Gardeners of high status also included the so-called 'gardeners of the flower offerings for the god', who were entrusted with the task of providing vegetables, plants and flowers for the altar in cult ceremonies. Garden designers were also held in high esteem and the job of laying out important gardens was sometimes given to official architects employed directly by the pharaohs, underlining the close link between monumental architecture and landscaped gardens. Senenmut and Amenhotep, son of Hapu, are just two such architects who designed the gardens associated with monuments of Hatshepsut and Amenhotep III. The chain of command from high-ranking officials to their underlings can be recognized in the preserved correspondence between Sennefer, the mayor of Thebes, and the gardener Baki. In a letter, Sennefer instructs the gardener to tend a garden in his care as follows: 'Don't keep it from being in proper order, but pick for me many plants, lotus blossoms and flowers to be made into bouquets that will be fit for presentation.'[1]

Unlike the administrators and designers of gardens, Egyptian texts make it very clear that the simple gardener or agricultural labourer had a hard life of toil (see p. 11). The gardener was responsible for sowing, harvesting and picking fruit from trees. Tending to vegetables was another simple but back-breaking task. The dry Egyptian climate made constant irrigation necessary, and if the gardens were near a source of water, the gardener could use a *shaduf* to irrigate them.[2] The *shaduf*, a device employed in Egypt and Assyria to lift water from a river or canal and direct it into irrigation channels, consisted of a long pole with a bucket on one end and a stone or lump of clay on the other as a

Detail of fig. 74 (p. 93)
Agricultural labourers prune trees in a scene on a Roman floor mosaic from St-Romain-en-Gal, France, c. AD *200–225.*

63 A shaduf *is used to draw water from a canal for the irrigation of garden plants in an Egyptian painting from the tomb of Ipui at Thebes,* c. *1240* BC.

counterpoise (fig. 63). However, if the gardens were at a distance from the river, well or canal, the gardener had to transport the water himself by shouldering a yoke with buckets or pots full of water. It was not until the Ptolemaic period that the *saqiya* (water wheel) and the Archimedes screw were introduced to Egypt – both these devices ensured a continuous flow of water and thus enabled perennial irrigation on a large scale (see fig. 58).

Far more information is available about gardeners and vintners in Egypt in the Hellenistic period. Under the Ptolemies, the Fayum marshes west of the Nile were drained by canals and dams and transformed into an extremely fertile agricultural zone in which huge orchards and vineyards flourished. All agriculturally worked land in Hellenistic Egypt was the property of the king, and even land that the king might 'give' to his important administrators remained, in reality, state property. A royal monopoly was held on certain products of the estates and taxes were paid on them. One of these taxes, the *apomoira*, entailed one-sixth of the annual harvest of fruit and wine from the gardens, *paradeisoi* and vineyards. The tax on orchards was payable in silver and that on vineyards in wine. The best known of these Ptolemaic estates were those given to and run by the chief financial adviser to the king, Apollonios.[3] Gardeners and vintners were employed on the estates, and all of these were Greek specialists with the necessary expertise to transform the Fayum into extensive plantations.

A great deal of correspondence between Apollonios and the estate personnel is preserved in papyrus documents (the so-called Zenon correspondence). Menon, a vintner in the third century BC, for example, had received a salary of three drachmas, but he lodged a complaint that his pay was too low. This was because he had not received the bonus of being allowed to pocket part of the proceeds from the sale of wine and vegetables from the estate. The vintners working for Apollonios normally received part of the profits and they were also supplied with tools and equipment. Centuries earlier, the specialists working in the palm groves in Mesopotamia had also had contractual agreements with the king, according to which they received a third of the harvest in payment.[4] Apollonios employed not only vintners but also gardeners in the Fayum. They too received a salary and were occasionally paid in kind; some of them also worked in the vineyards, where they, like the vintners, may have shared in the profits. In the papyri, the tasks carried out by Apollonios's gardeners included the growing, care and transport of seedlings and young trees. The latter sometimes were brought from the estates in Memphis to the orchards in Philadelphia.

The tasks of a gardener are outlined in a unique agreement, possibly dating to the late third century AD, written in ink on a jar from Medinet Habu in Egypt.[5] In it, Talames, the owner of a garden, contractually regulates the responsibilities and rights of Peftumont, the gardener. Peftumont is required to water the garden and maintain its irrigation channels; to make four baskets of palm fibre for earth; to protect the garden against sparrows and crows and to complete his work at the end of each day. At this time, he is also to hand over his own excrement for inspection, so that Talames may 'probe it with a stalk' to ensure the gardener had not eaten the produce of the garden. Alternative

64 Basil in the illustrated herbal of Dioscorides copied in AD *512 in Constantinople.*

methods of payment for his labour are discussed, and he could choose to have his wages paid in wheat, gold or bronze.

Work involving trees is one of the tasks specified as a gardener's responsibility in Greek and Roman historical accounts. King Eumenes II of Pergamon employed gardeners to plant new trees and nurse damaged ones back to health in the grove of the Nikephorion, following an attack on the city by Philip V of Macedon. On Philip's orders, his army had cut down many of the trees in the sanctuary in 201 BC (Polybios 16.1.6).

Watering and weeding were also natural tasks of the gardener. In Greek agricultural literature, water from springs, rivers and canals was differentiated, each source of water having different properties and qualities. Onions were thought to be cultivated best with fresh, cold water, not with sluggish water from irrigation ditches. It also specified the amount of water required: garden herbs required watering twice a day, although basil needed a third daily watering (fig. 64). Irrigation ditches and pipes were common in Roman gardens, and the underground system of channels and sluices, which watered the plants 'from below', in the Piazza d'Oro at Hadrian's villa in Tivoli was a particularly sophisticated labour-saving device (fig. 65).[6] Underlying the garden soil, a system of rock-cut trenches directed pumped water through the garden, with the garden soil above being irrigated by capillary action. When the soil was moist enough, a sluice was shut and the flow of water interrupted for a period of time. The entire system was designed to mechanize irrigation and to restrict manual labour to the handling of a sluice. In most Roman gardens, however, water was brought to the plants by the gardener in buckets or containers. Columella recommended that footpaths accessed the vegetable-beds and flower-beds so that the gardeners did not tread on the plants (*De re rustica* 2.10.26).

The earliest record of the profession of gardener in Greece is preserved in a list of resident foreigners, that is non-Athenians, who received state honours for services rendered to Athens in 404 BC.[7] Since resident foreigners were not allowed to own land in Athens, this gardener was probably a hired employee on someone's farm or estate. In the same list, other professions such as farmer and vintner are recorded. Theophrastos mentions gardening as a profession at the end of the fourth century BC, and stresses how important gardeners are for the proper care of vegetable gardens (*Characters* 20.9). Nevertheless, there are very few references to gardeners, probably because the profession was still rather unusual in Classical Greece, where most farmers were small landowners who did their own agricultural work. Nor were hired employees of high social status in Classical Greece. Aristotle considered it undignified for any man to earn a living by working for another (*Politics* 1367a), and the larger landowners made ample use of slaves to work their estates, rather than hired personnel. On the vast royal estates and in the palace gardens of the Hellenistic monarchs, however, specialists were needed for their expertise, even if their status was not much better than it had been in the Classical period. By the Roman period, the profession had become firmly established, and gardeners formed professional guilds.

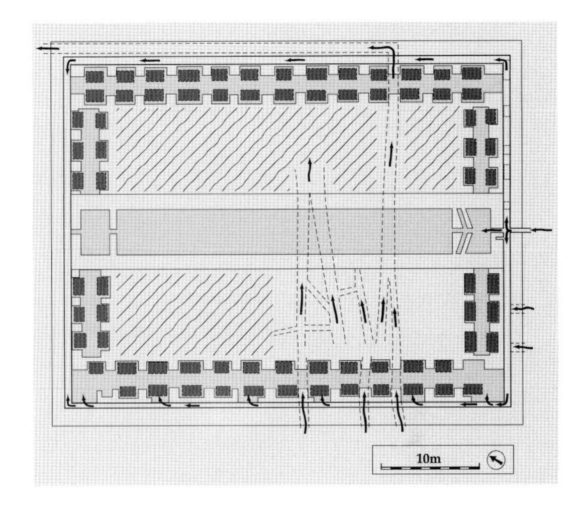

65 Water (black arrows) flowing through underground channels and sluices irrigated the garden and plantings in pits around the edges (dark grey boxes), and fed the long central pool (light grey) in the Piazza d'Oro at Hadrian's Villa at Tivoli, c. AD 118–134. Unexcavated areas are marked with wavy lines.

One of the new specialist branches of the profession in the Roman period was the garden artist or topiary specialist. This speciality is first mentioned in a letter from Cicero to his brother in 54 BC, reporting on the work being done by a *topiarius* in his brother's garden (*Q. Frat.* 3.1.5). The *topiarius* created garden landscapes, making use of artistically arranged ivy and planting trees and bushes in symmetrical patterns and circles (Pliny, *Natural History* 12.22.22). By the late first century BC, these gardens came to include trees clipped in various shapes, including hunting scenes and whole fleets of ships (Pliny, *Natural History* 12.6.13, 16.31.76, 16.60.140). At Pliny's Tuscan villa, 'there are box shrubs clipped into innumerable shapes, some being letters

66 Clipped box hedges in the central garden at Fishbourne Roman Palace, c. AD *75, replanted in 1968 according to the ancient planting patterns.*

which spell the gardener's name or his master's . . . farther off are acanthuses with their flexible glossy leaves, then more box figures and names' (*Letters* 5.6.35-36). That the *topiarius* was held in high esteem by the garden-loving Romans is reflected in contemporary literary sources by the naming of the man who invented this form of garden art: Gaius Matius, who lived at the time of Emperor Augustus (reigned 27 BC–AD 14). A sense of pride in the profession is conveyed in the epitaphs on gravestones in which the deceased chose to record his professional status as a *topiarius. Topiarii* perhaps practised their art outside Italy, and we can assume that the personnel who designed and planted the garden with its formal hedges at Fishbourne in Britain were immigrant gardeners (fig. 66). Not only was topiary widespread during the reign of Augustus, but landscaped ornamental gardens also became a popular motif in Roman wall painting (fig. 67). Favourite plants in these gardens were ivy, box, laurel, cypress, myrtle, acanthus, dwarf plane trees and rosemary.

67 A room in Livia's villa at Prima Porta near Rome was painted as a garden with a profusion of plants, early first century AD.

Practical literature on farming and agriculture was compiled at least as early as the fourth century BC in Greece. The Greek philosopher Theophrastos published two books on plants and agriculture known as the *History of Plants*

and the *Enquiry on Plants*, which contained information and practical tips that he had gathered from landowners. By the time the Roman author Varro wrote his treatise on farming in 37 BC, he could cite over fifty different authors from whose works he had borrowed (*Rerum Rusticarum* 1.8). These books give plenty of advice such as recommending the size of planting pits for fruit trees and vines (Xenophon, *Oikonomikos* 19.3–5). In Alexandria, the law-makers actually decreed the distance olive, fig and fruit trees should be planted from the property boundary in order to protect the neighbour's property from being damaged by falling trees and fruit.[8] Theophrastos also included practical tips on the treatment of flowers to induce them to bloom all year long (*History of Plants* 6.8.1–4). This is related to the cultivation of flowers for commercial and cultic purposes, since flower wreaths and garlands were frequently used in religious festivals (fig. 68). Favourite flowers for these were white violets, narcissus, carnations, anemones, hyacinths, roses, lilies, irises and crocuses. Practical literature specifically concerning gardens, rather than agriculture and crop husbandry, however, does not appear to pre-date the Roman period. Sabinus Tiro, a Roman of the late first century BC, is credited with having written a book entitled *On Gardening* (Pliny, *Natural History* 19.57.177). The book is now lost, but, based on Pliny's remarks, it contained information on plants and plant pests and practical tips such as how to rid the garden of ants. Two of his recipes for this were to plug the mouth of the ant-holes with ash or sea slime, or to plant the heliotrope plant in the garden (fig. 69).

68 A woman kneels at an altar with a wreath of leaves or flowers in a scene on a red figure cup from Athens, fifth century BC.

Roman writers on agriculture described a particular method for planting new trees that involved clay plant pots. Cato (*On Agriculture* 51–2, 133) and Pliny (*Natural History* 17.21.97–8, 17.11.64), for example, recommended various ways of starting trees. These methods were: (a) pulling down a shoot of the tree, covering it with soil and letting it root in a pot; (b) air layering, which involved leaving a pot, through which a shoot was fed, suspended on the tree until the shoot had taken root; and (c) planting seeds in pots. Cato also suggested that the pot, in which the root of a young tree was growing, be broken before the pot was placed in the ground to allow the roots to continue to grow unhindered:

> When you wish to layer more carefully you should use pots or baskets with holes in them, and these should be planted with the scion in the trench. To make them take root while on the tree, make a hole in the bottom of the pot or basket and push the branch which you wish to root through it . . . When it is two years old, cut off the branch below the basket; cut the basket down the side and through the bottom, or, if it is a pot, break it, and plant the branch in the trench with the basket or pot.
>
> (Cato, *On Agriculture* 52)

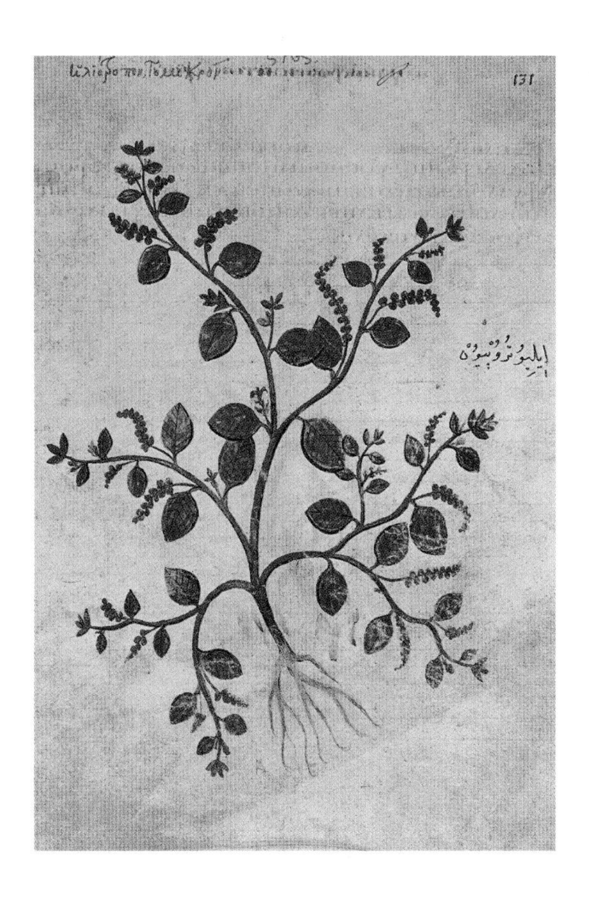

69 The heliotrope plant in the illustrated herbal of Dioscorides copied in AD 512 in Constantinople.

Cato and Pliny named vines, lemon trees, pomegranates, quinces, nut trees, rosemary and oleanders specifically in this connection.

This practice can be related directly to archaeological discoveries of plant pots in many Roman gardens and not only in Italy. Most of the plant pots found in Pompeian gardens are of a tall and slightly ovoid shape, with one hole in the bottom and three in the lower body of the pot.[9] The Pompeian pots are most frequently found in planting pits in house gardens, but one has also been excavated recently in the precinct of the temple of Venus (fig. 70). Just as Cato suggested, this pot had clearly been given a blow with an instrument to crack it before it had been placed in the pit. At the villa of Poppaea at Oplontis, excavations revealed that lemon trees had started off in clay pots.[10] There were several advantages to using clay containers for young trees. Firstly, the plants could be raised in larger numbers in plant nurseries or in existing orchards; secondly, the porous clay of the pot retained moisture, so that the young roots would not dry out so quickly once watered; and thirdly, trees and shrubs could be transported over longer distances to the gardens for which they were intended.

70 Terracotta planting pot from the temple of Venus in Pompeii, first century BC. *Three holes pierced in the lower body of the pot and one in the base allowed root growth.*

71 Plan of the courtyard garden of the Roman villa at Richebourg, France, c. first century AD. *The black dots represent the position of planting pots in the soil.*

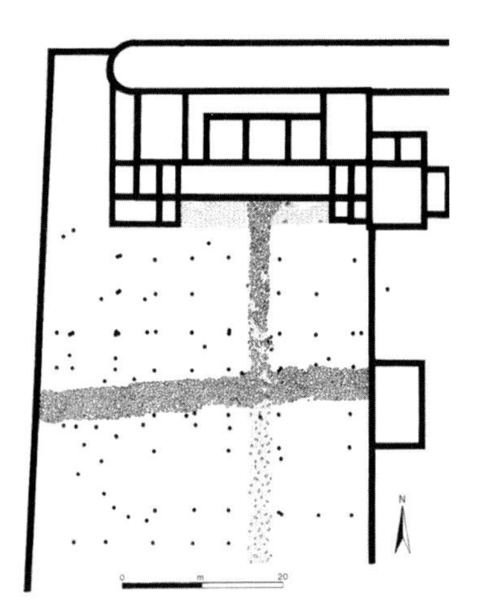

Importing trees and plants from the eastern Mediterranean to Italy reached its peak in the first centuries BC and AD. Pliny referred to the introduction of citron trees 'in earthenware pots provided with breathing holes for the roots' (*Natural History* 12.7.16), and he also listed the cherry, the peach, the apricot and the pistachio tree as newcomers to Italy. The export of Mediterranean plants to the western and northern Roman provinces may also have been facilitated by transporting them in pots. Certainly trees and plants can be found in many European provinces which were not indigenous to the area prior to the Roman conquest (see Chapter 7).

Since their discovery in Pompeii, Roman plant pots have been found at several sites outside Italy, including Jericho in Palestine, Carthage in Tunisia, Conimbriga in Portugal, Fishbourne and Eccles in Britain, and Lyon, St-Romain-en-Gal, Nîmes and Richebourg in France.[11] At Richebourg, more than one hundred plant pots have been excavated in the courtyard in front of a rural villa of the first century AD (fig. 71). These were spaced at fairly regular intervals around the two gravel paths, and it appears the plantings were formally arranged. Although the plant pots from the sites are not identical in shape, they all have holes in the body and base of the pots. It is rarely certain whether or not these pots (with plants in them) were imported or whether they were produced locally, but pots from Eccles in Kent were definitely produced locally in the middle of the first century AD.[12] That this very Roman practice was introduced to Britain so soon after the conquest of the island in AD 43 clearly shows how essential and important the garden was to the Romans, even when they went abroad.

The recent discovery of plant pots in Rome can even help to shed light on the landscaping of a large monument in the north-east sector of the Palatine hill that was known only from written sources. Excavations conducted by the French have revealed the foundation walls of the temple of the oriental god Elagabulus, erected by the emperor Elagabalus in the early third century AD and later re-consecrated by his successor Severus Alexander between AD 222 and 235 as the temple of Jupiter Ultor. This temple, the Heliogabalium, stood in a paved courtyard that was surrounded by porticoes on all four sides. The paving was interrupted by four rectangular planting beds in which rows of halved amphorae were arranged.[13] Halved amphorae, re-used as planters, have also been found in some Pompeian gardens, and in the garden beds flanking the Canopus at Hadrian's villa in Tivoli.[14] The modest pots from the Palatine hill thus confirm that this large monument to Elagabalus in the heart of Rome was designed as a garden.

The excavators originally identified the building remains and the plantings as the sanctuary of Adonis, or Adonaea, known from its depiction on the marble plan of Rome (fig. 72).[15] This plan shows a T-shaped edifice with numerous drilled dots and lines that are usually interpreted as plantings, perhaps trees and shrubs. This interpretation is also based on the knowledge that, in Classical Athens, potted plants were part of the ritual practised in the cult of Adonis, the youth whose premature death was mourned by Aphrodite, but who returned to life

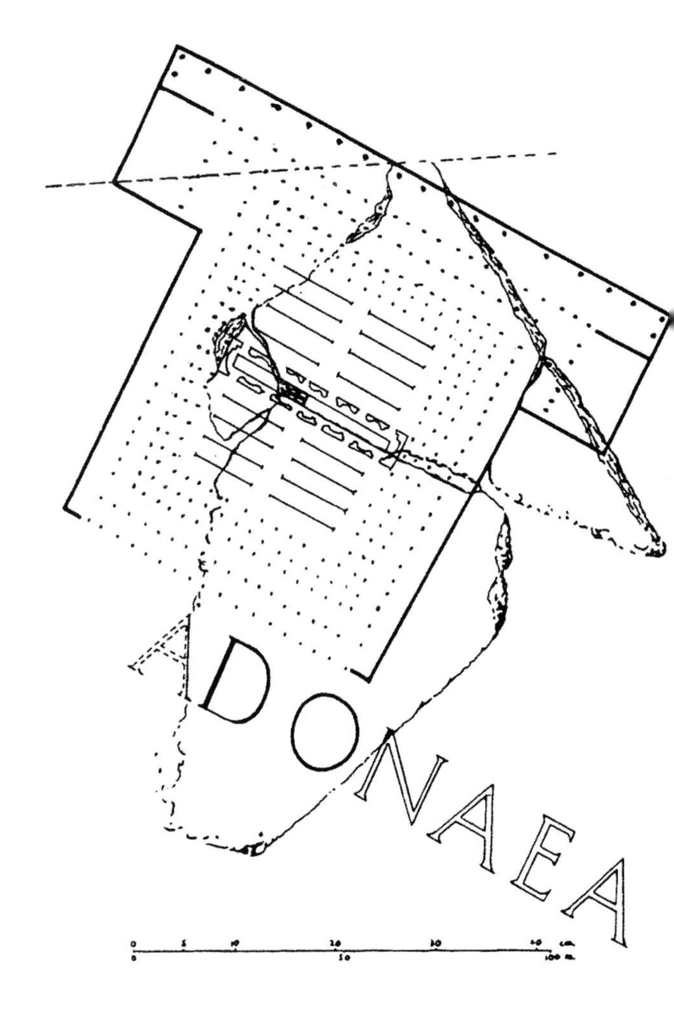

72 Temple of Adonis in Rome on a marble plan, early third century AD. The dots possibly represent the position of plants or shrubs grown in pots.

for six months of the year.[16] Until the Adonaea is located archaeologically, however, it cannot be determined whether the dots on the plan really represent potted plants.

Although used widely in the Roman period, plant pots used to propagate and transport plants were not invented by the Romans. Theophrastos, in his *History of Plants* (4.4.3), referred to the use of perforated clay pots for starting cedars and palm trees in Persia and Media in the fourth century BC. In the third century, the tyrant of Syracuse, Hieron II, grew plants and various trees in pots on his luxury ship (Athenaeus, *Deipnosophists* 5.207d–e). Archaeological evidence for clay plant pots in the soil was retrieved in the 1930s during the excavations of the precinct of the temple of Hephaistos on the edge of the Athenian agora (fig. 73).[17] Clay pots were found broken and embedded in the soil that filled square cuttings in the rock at a depth of about 50 centimetres. By digging or cutting pits and filling them with garden soil, even the most unfavourable ground, such as rock or gravel, could be made suitable for plantings. It has been calculated that about sixty shrubs or small trees encircled the temple of Hephaistos, at least until the first century AD, when the irrigation system supplying the garden was abandoned. These pots manufactured for placing in the soil were larger than those made for use above ground. The latter type of pots, some as small as 8 centimetres in height, with a hole in the base or body of the vessel, are known from houses of the fourth century BC in the Greek city of Olynthos.[18] It is likely that such plant pots contained herbs to use for cooking and stood in the paved courtyard of the house.

73 Rock-cut pit containing a terracotta plant pot at the temple of Hephaistos in Athens, c. third–first century BC.

To perform their tasks, gardeners needed a variety of tools. Many of the iron objects found in excavations are so corroded that their function can no longer be determined, and it is not always clear whether they were garden-specific or generic tools for all kinds of agricultural work (fig. 74).[19] Hoes, rakes, sickles, axes and ploughs are occasionally depicted in Egyptian wall

74 Agricultural labourers prune trees with a curved pruning knife in a scene on a Roman floor mosaic from St-Romain-en-Gal, France, c. AD *200–225.*

paintings of work in the fields rather than in the garden. This does not mean that these implements were not used in horticulture; the gardens in Egyptian art are usually idealized ones in which back-breaking labour does not take place. Hoes, sickles, pruning knives and axes have been found at Classical and Hellenistic Greek sites such as Olynthos, Corinth and Priene.[20] Hoes, particularly the double-pronged variety, were used to break up the often hard and stony soil in Greece, and their employment in vineyards is recorded in literary sources (Menander, *Georgos* 65). A reference to their use in the garden,

however, indicates that double-pronged hoes were suitable for both agricultural and horticultural work (Hyperides, *Against Demosthenes*, Frag. 6).

Preserved hoes and hatchets found in Pompeii for loosening the soil and cleaning weeds and scrubs from gardens are probably what ancient sources refer to as the *sarculum* and *dolabella*.[21] Sickles, pruning knives and other agricultural tools are occasionally represented as tools of the trade on Roman gravestones, and miniature tools can be found in late Roman graves.[22] Roman spades were often made of wood, but the blade was reinforced with a sheath of iron to form a sharp cutting edge. These iron blades have been found at many Roman sites in Europe.[23] Any of these agricultural implements could have been used to work the soil of a courtyard garden, or a nearby vegetable garden, or plots in the surrounding fields.

Wooden gardening implements rarely survive, so it was a surprise to find the impression left by a wooden ladder lying on the ground in the peristyle garden in the house of Polybius in Pompeii in 1973.[24] The ladder was 8 metres long and narrowest at the top, just the kind of ladder used today to pick fruit from tall, densely foliated fruit trees. Wooden stakes and posts were frequently employed to prop up plants and construct a trellis or arbour, although all that normally survives of these are the voids in the soil left by wood that has since rotted. Careful excavation of small holes and carbonized wood in these holes in the courtyard of the so-called Casa dei Casti Amanti in Pompeii revealed that trellises on the edges of the garden were made from stakes of the giant reed with crossed canes of the common reed in between them.[25] A hollow reed is depicted as a support for a rose bush in the Roman wall paintings from the Casa del Bracciale d'Oro in Pompeii (fig. 75), and canes and reeds are also mentioned as supports for vines in a garden in Philadelphia in Hellenistic Egypt.[26]

Various different types of plants and trees have been referred to in this book. A more detailed examination of these, and the surviving evidence for them, is the subject of the next chapter. Ancient gardeners fertilized the soil to improve the quality of the growing medium of their plants. All Greek and Roman books on agriculture contain helpful information on how to enhance the fertility of the soil. The next chapter also explores how environmental archaeology and soil analysis today can help reconstruct these horticultural practices.

75 Detail of a Roman garden painting in the Casa del Bracciale d'Oro in Pompeii showing a rose supported on a reed cane, first century AD.

CHAPTER SEVEN

Plants of the ancient world

Since at least the second millennium BC, plants had been transported and introduced to new areas. Recent archaeobotanical evidence suggests, for example, that the olive was introduced to Egypt possibly as early as the middle of the eighteenth century BC, and by the later second millennium it had become a staple agricultural product in Greece (fig. 76).[1] In the early fifteenth century BC, Queen Hatshepsut had myrrh trees, roots and all, brought to Egypt from upland Ethiopia (see p. 42), and in the ninth and eighth centuries BC Mesopotamian kings introduced plants such as the cedar, the oak and rare fruit trees, as well as spices and other vegetation from Media and Anatolia to their parks and gardens (see pp. 43–4). The elder Dionysius, tyrant of Sicily, imported plane trees to Rhegium to adorn his palaces around 390 BC, and in the Hellenistic period the Greek monarchs introduced Arabian frankincense plants to Syria and Egypt (Pliny, *Natural History* 12.31.56–7).

From the first century BC, the Romans introduced species of trees from the eastern Mediterranean and the Near East to Italy. Books 12 to 15 of Pliny's *Natural History* list many trees that were imported into Italy, such as the cherry, peach, apricot, pistachio, damson plum, medlar and citron. By the end of the first century AD, these trees had become established domesticates in Roman gardens in Italy. From there, the Romans in turn transported plants and trees to the areas they had conquered in western Europe. According to Pliny (*Natural History* 15.102), cherry trees were brought from Pontus on the Black Sea to Italy in 74 BC, and 'in 120 years they have crossed the ocean and got as far as Britain'. Southern European vegetation was not always suited to more northerly climates, however, as shown by the difficulties encountered by a modern replanting of an Italian cypress tree at Fishbourne. Vegetables, fruit

Detail of fig. 83 (p. 105)
Roman garden painting in a room in the Casa del Bracciale d'Oro in Pompeii, first century AD.

76 *Olives, a staple of ancient Greek agriculture, being harvested in a scene on a black figure amphora from Athens, sixth century* BC.

and herbs from southern Europe, Asia and Africa, such as the cucumber, coriander, fennel, dill and the mulberry, were introduced to Britain in the Roman period. In Germany, figs, chickpeas, plums and peaches, for example, were brought by the Romans. The introduction of exotic plants and trees to newly conquered areas continued under the Arabs who, in the eighth and ninth centuries AD, brought Syrian pomegranates and Persian oranges to Andalusia and Sicily (fig. 77).[2]

It is often difficult, however, to determine whether whole plants were actually exported to western and northern Europe for cultivation in gardens in the Roman period or whether only the fruits and seeds were brought there for use in cooking and in medicine. This uncertainty has to do with the fact that the plant remains tend to be only fragmentarily preserved.[3] Unfortunately they are not normally found in garden soils since the soils have been repeatedly worked – by humans, animals, worms and burrowing organisms – and are well aerated so that plant remains usually decay rapidly. Instead, macroscopic fossil traces of plants are found in middens, pit fills, wells, cesspits and water-logged deposits, as well as in containers, such as amphorae, in which foodstuffs were shipped. Carbonized plant remains can also provide evidence for plants, either imported or locally grown varieties. At Pompeii, for example, carbonized hazel nuts, figs, grapes, almonds and broad beans were found in a market garden, as were bits of carbonized roots of the hazel and plane tree.[4] Whole carbonized fruits are much rarer at western and northern European sites. However, charred apples have been retrieved from the cellar of a Roman farmhouse at Lörrach-Brombach in south-west Germany and burnt imported fruits and nuts, such as figs, dates, almonds, hazelnuts and walnuts, have been found in a cremation burial in Xanten on the lower Rhine.[5] The recent discovery of several burnt timber barrels containing pomegranates in a storage building of the early first century AD at Vindonissa in Switzerland is

77 *Oranges, pomegranates and other trees in the Islamic garden (Bagh-i Wafa) of Babur in Kabul, depicted in the illustrated memoirs of Babur, sixteenth century* AD.

میان غرب و جنوب باغ حوض ده در دهی است اطراف و تمام
درختهای نارنج است درختهای انار هم هست گرداگرد حوض تمام
سبزه زار است جای عین باغ همین است در وقت زرد شدن
نارنجها بسیار خوب می نماید خیلی باغ خوبی طرح شده و طرف

unique, as it is the only Roman site north of the Alps where the remains of this Mediterranean fruit have been found.[6] The carbonized material from these barrels also included other Mediterranean fruit such as olives, pistachios, peaches, cherries and dates.

In the dry climate of Egypt, botanical remains have survived intact in a number of tombs. The tomb of Tutankhamun, for example, contained a wealth of local flora, including papyrus, olive leaves, cornflowers, water-lily petals, wild celery leaves, pomegranate leaves (fig. 78) and persea fruits, woven into garlands and necklaces.[7] Extreme conditions such as these, however, are rare and archaeologists must normally resort to other methods to retrieve botanical remains.

78 Flowers and developing fruit of a pomegranate tree.

An effective method for recovering organic material from archaeological deposits is that of flotation, which involves the use of water, either in buckets or in a machine, to separate organic material from soil samples collected during excavation. Charred remains of plants will float to the surface and can be caught in fine mesh sieves; desiccated remains can also be retrieved in this way, although equally these can be detected by dry sieving. If the remains are large enough to be visible to the naked eye, for example cherry pits or walnut shells, they also can be hand-selected from soils during excavation. Once the plant remains have been collected using any of these methods, they are then sorted under a microscope and identified using modern comparative botanical material.

Samples of dung excavated in animal pens at the Workmen's Village at Amarna in Egypt were analysed in this way and were found to contain many botanical remains, indicating that the animals were fed largely on cereals but also on other plant material discarded on rubbish heaps.[8] Identified material included cereals, carbonized seeds and nuts, kernels from the dom palm and date palm, as well as pomegranate seeds, grape pips, sycomore fig seeds, persea kernels, olive pits and watermelon seeds. Analysing this material gave an insight into the diet of humans and animals in Amarna and also into the crops grown in the vicinity, since many of these foodstuffs were probably brought into the city from the outlying farms and estates.

Microscopic remains of vegetation, in the form of pollen, can also be detected in soils in favourable conditions. In the case of these remains, it is clear that the plants from which the pollen came were cultivated locally. The clusters of pollen of common mallow found in deposits of Roman sewage at the fort at Bearsden on the Antonine Wall show that this medicinal plant was grown locally in the second century AD.[9]

Some foodstuffs and plants were not shipped over great distances; the macroscopic remains of a root and part of a stalk of a cabbage recovered at Chesterholm fort on Hadrian's Wall suggest vegetable gardening in the vicinity.[10] Equally, the remains of the delicate leaf epidermis of leeks, preserved in water-logged deposits in a cesspit at York, confirm that this plant was cultivated locally for culinary use.[11] The fruit of the mulberry tree has delicate, perishable fruit that is not normally transported far, so the mulberry fruit found in London, Silchester and York was probably grown in nearby orchards and gardens.[12]

79 Box hedges clipped in geometric patterns in the garden of the Roman palace at Fishbourne.

Other macroscopic remains frequently recorded at Roman sites in Europe are clippings of box, and this shrub, as well as holly and the Portugese laurel, may represent a Roman import that was subsequently grown in European gardens as ornamental trees and shrubs.[13] It was probably for box hedges that the bedding trenches, arranged in geometric designs, were dug into the gravel subsoil in the garden at Fishbourne in Sussex (fig. 79).

Seeds such as dill, celery and coriander were probably imported in large quantities for use as seasoning in cooking or for medicinal purposes, and pulses such as peas were probably imported as bulk foodstuffs, possibly in

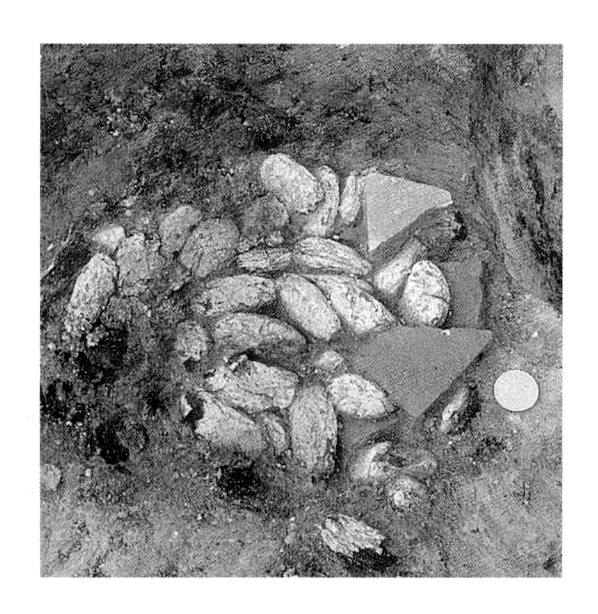

80 Carbonized dates from the Mediterranean, found in the destruction debris from the Boudican revolt in Colchester in Essex, AD 60.

dried form. The preservation of such remains in human excrement points to their consumption, but this does not reveal whether they were imported or cultivated or for what reason they were eaten. It is fairly certain, however, that the fig seeds found at sites in the western and northern European provinces represent imports of dried fruit from the Mediterranean, rather than fruit from locally grown trees. Fig seeds, coriander seeds, grape pips and peppercorns found in deposits of rubbish at the fort at Oberaden in Germany are clearly the remains of food that was imported to this region east of the Rhine when the Roman army was campaigning against Germanic tribes in the late first century BC. None of them, apart from the grape in its wild form, grew north of the Alps.[14] Other imported Mediterranean fruit in the north include the carbonized dates found on the floor of a reused barracks block at Colchester that was destroyed in the Boudican revolt in AD 60 (fig. 80).[15] In general, exotic fruit and plants were imported from the Mediterranean world to northern Europe to feed the thousands of Roman soldiers stationed on the frontier in the early first century AD. By the second half of that century, and especially in the second century, such imports became more widely available to the native civilian population, both as traded items and some of them, such as dill, garlic and gourds, as species that were grown locally.[16]

Certain plants were thought to have medicinal properties. The root of the black hellebore was considered a medicinal plant stuff (fig. 81), the best coming from Mount Helicon which was generally 'well supplied with drugs' (Theophrastos, *History of Plants* 9.10.3). Cato recommended cabbage, 'which surpasses all other vegetables', either cooked or raw, as an aid to digestion and as a laxative (Cato, *De Agri Cultura* 156). Walnuts were supposed to aid in expelling tapeworms, and cucumber roots, dried and compounded with resin, were taken to heal scabies, psoriasis and ringworm (Pliny, *Natural History* 20.3, 23.148).

Although representations of plants and trees in ancient art may represent an idealized version of a garden environment, they give us some idea of the types of vegetation that existed and the contexts in which they grew. In Egyptian paintings sycomores, date palms, tamarisks and fig trees appear frequently, as do papyrus and lotus in pools and channels. These depictions are stylized, but nonetheless recognizable. That these plants and trees were truly elements of gardens and orchards is confirmed by archaeological evidence: examples include the preserved botanical remains of roots, bark and leaves of tamarisks and sycomores in the planting pits at the mortuary temple of Mentuhotep II at Deir el-Bahari, the desiccated papyrus stems in the pools of the mortuary temple of Hatshepsut and the wreaths of dried flowers found in Egyptian tombs. Palm trees, the tree *par excellence* of Mesopotamia, are both represented in Assyrian reliefs and recorded in written records of the region. Moreover, the relief panels adorning the rooms of the palace of Sennacherib at Nineveh in the early seventh century BC reveal an acute interest not only in the portrayal of cultivated gardens but also in the depiction of all manner of trees and vegetation of the natural environment (fig. 82). In Classical Greek art gardens are rarely represented since garden painting did not exist as a genre.

81 The hellebore plant in the illustrated herbal of Dioscorides copied in AD 512 in Constantinople.

A garden with a profusion of plants and trees decorates the walls of the so-called Villa of Livia at Prima Porta (Rome) dating to the late first century BC (see fig. 67).[17] In this garden, conifers and oak trees growing on a lawn in the foreground are separated by a fence from quince and pomegranate trees, palms, laurels, myrtles, oleanders, roses, chrysanthemums and a wide variety of other flowers. In the garden paintings from the Casa del Bracciale d'Oro in Pompeii, dwarf planes, laurels and viburnum grow behind roses, strawberry trees (*arbutus*), oleanders, date palms, ivy, poppies, camomile, violets, morning glories and madonna lilies (fig. 83).[18] These garden paintings show a

82 Pine trees lining a river on a wall relief from the south-west palace of Sennacherib at Nineveh, between 700 and 681 BC.

83 Roman garden painting in a room in the Casa del Bracciale d'Oro in Pompeii showing a variety of shrubs, trees and flowers, first century AD.

variety of trees, bushes and flowers, carefully arranged so that all plants are visible. The smallest and most delicate plants grow profusely in the foreground, whilst the taller trees dominate the background. These are artistic garden compositions, and not only is the vegetation arranged in an idealized manner but the marble herms and flowing fountains are also placed symmetrically and within full view in the foreground for maximum effect. Not all the vegetation could have flowered and fruited at the same time of year. However, even if these paintings are idealized compositions with a still-life quality to them, the naturalistic depiction of the flora is striking, and it shows that the painters had observed the garden plants very closely. It is also worth noting that the many accurately depicted birds in these gardens were inserted into the picture, probably by specialist bird artists, after the painting of the vegetation had been completed by the garden painters.

Garden paintings such as these decorated the walls of entire rooms, transforming an indoor space into a virtual garden. In other Roman houses,

however, garden paintings can be found in the open courtyard, especially modestly sized ones. By painting gardens on the interior walls of the courtyard, the planted area could be 'enlarged' with artistic illusion (fig. 84). House owners of limited means could thus attempt to 'compete' with those who could afford a really large and luxurious garden as a symbol of their status and wealth.

The tradition of adorning the internal walls of buildings with paintings of gardens and orchards was continued in the late Roman period in Egypt and developed to include gardens depicted on large textile tapestries. A number of fragmentary wall tapestries of linen and wool from the fifth century AD survive, in which a row of trees is woven in brilliant colours, the most common types being lotus, pomegranate, fig, apple and peach (fig. 85).[19] These magnificent tapestries depict fruit-bearing and flowering trees that probably grew in contemporary gardens and orchards, but they may also have alluded to the heavenly gardens of paradise. In sixth-century floor mosaics in churches in Jordan the association of rows of fruit trees and animals with the garden of Eden is clearer, especially when combined with the four rivers of paradise (see p. 45 and fig. 35).

Fruit, vegetables and wheat were also represented in sculpture of the Roman period. The many Roman votive stone altars dedicated to the *Matronae* (ancestral mother goddesses) worshipped on the lower Rhine in Germany show the triad of *Matronae* with a basket of fruit on their laps. Offerings to these goddesses, who ensured fertility and protected the home, were the fruits of the field and the orchard, but no specimens of real fruit have survived in

84 (Previous page) *The lack of a sizeable garden was redressed by the painting of a garden on the wall (right) in the courtyard of the House of Neptune and Amphitrite in Herculaneum, first century AD.*

85 (Above) *Fragment of a larger textile tapestry from Coptic Egypt depicting lotus trees, early fifth century AD.*

excavated sanctuaries. Some of the *Matronae* altars, however, have pieces of fruit carved in stone on the top of the altar, sometimes in a *patera* or offering dish used in the ritual of worship (fig. 86). The products of local orchards and vineyards of the second and third centuries AD, such as apples, pears and bunches of grapes, can clearly be seen.

86 Offerings of fruit, in this case a pear, were carved on the top of stone altars dedicated to the Matronae *near Cologne on the Rhine, third century* AD.

Identifying soils deposited for growing garden plants can be difficult, especially since we commonly tend to think of garden soils as fine, dark in colour and humic. Three of my own excavations in Roman and medieval gardens served to dispel this perception of 'classic' garden soils and instruct us that plants, trees and shrubs will grow in soil that can look very different from the fine loam and compost that we buy in modern garden centres. At the temple of Venus in Pompeii, the soil in the pit containing a planting pot for a shrub was light brown and full of pottery and fragments of building material. The garden soils of the early fourth century in the possibly lawned courtyard of Chedworth Roman villa in Gloucestershire ranged from light brown to ochre and blackish brown in colour, and they contained pottery sherds, animal bone, shells and fragments of building material. The soil in the garden of the medieval Augustinian monastery in Heidelberg was indeed a dark, blackish brown colour, but it included numerous pottery sherds, glass, stones and animal bones. Some members of my excavation team flatly refused to believe that so much chunky rubbish would have been mixed with the soil in a proper garden. This medieval garden appeared to have been enriched with human excrement, kitchen waste and discarded domestic rubbish extracted from the latrine located immediately adjacent to the garden.[20]

The consistency and make-up of the soils depended on the crop being cultivated. Trees, shrubs and grass do not need the intense use of fertilizer nor the soil to be worked and turned repeatedly, whilst vegetable- and flower-beds are more likely to require fertilizer or the importation of finer topsoil to improve the growing conditions. Roman agricultural writers differentiated between types of land, that for vineyards, 'for a watered garden', for osier beds, for olives, for meadows, for grain fields, for woodlands, for orchards and for a mast grove (Varro, *Rerum Rusticarum* 1.7.9).

The enrichment of garden soils with manure was one of the contractual regulations of the leased temple estate of Zeus Temenites on the Greek island of Amorgos in the Classical period.[21] In the contract, the lessee was obliged to buy 150 baskets of manure for the vineyard and fig orchard from the temple managers; presumably this manure came from the herds owned by the temple. In urban houses at Olynthos, for example, the manure (*kopron*) from the animals kept in the courtyard was collected in a small walled enclosure next to the door of the house.[22] Both literary sources and inscriptions indicate that this manure came from animals and humans and was collected by private firms or people organized by the city administrators. The waste was brought to a place outside the city walls, where this 'night soil' was probably sold to the owners of suburban garden plots. The transportation of manure on a pallet to gardens or fields is represented on a Roman floor mosaic from St Romain-en-Gal (fig. 87). *Kopron* could also consist of cuttings and waste from plants and grasses (Xenophon, *Oikonomikos* 16.12). Greek and Roman agricultural writers suggested a variety of fertilizers, such as pigeon dung for meadows, gardens and field crops, but also compost of straw, lupines, chaff, bean stalks, husks, ilex, oak leaves and animal manure (Cato, *De Agri Cultura* 37.1–2, 39.1). The use of fresh human dung in making liquid manure was also recommended in gardening (Theophrastos, *History of Plants* 7.5.1). Another method of enriching soil was to mix topsoil with chalk to reduce acidity, as at Fishbourne.

In general, human and animal manure, the material from dredged ditches and ponds and kitchen refuse, to name but a few things, were gathered for compost and spread over fields and in gardens. Pottery sherds from broken crockery are common artefacts in cultivated soils, suggesting that waste from the kitchen was a component of compost. This applies no less to Roman gardens at Herod's palace in Jericho than to the plant-beds flanking the path in the courtyard at Frocester Court villa in Gloucestershire (fig. 88).[23] In the latter case, the soil of the plant-beds was mixed with 'potsherds, broken bones and jewellery, especially hairpins', although the hairpins should not be interpreted as an indication that the beds were tended by the ladies 'who took an interest in the flowers'.[24]

87 Agricultural labourers transport manure to fertilize a garden in a scene on a Roman floor mosaic from St-Romain-en-Gal, France, c. AD 200–225.

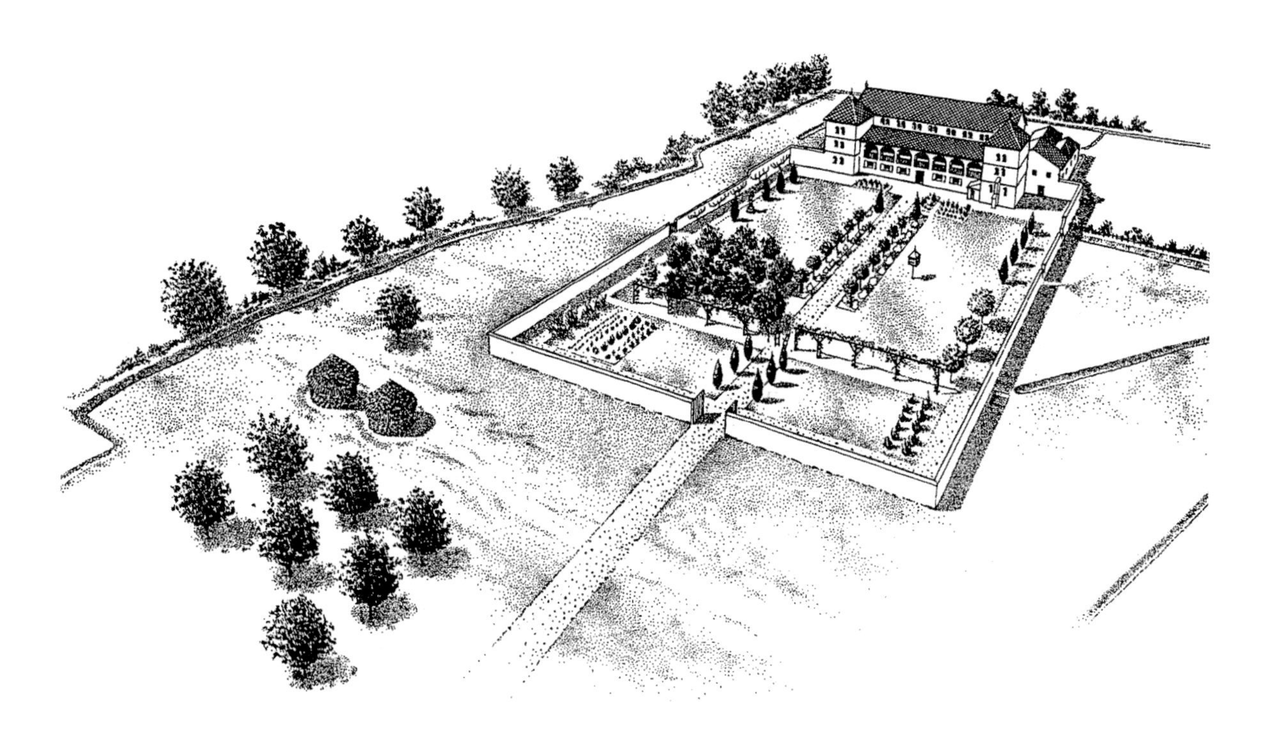

88 Reconstruction of the Roman villa and its gardens at Frocester in Gloucestershire, fourth century AD.

Other artefacts found in soils, such as molluscs, can be helpful in reconstructing the habitat in the garden.[25] Some snails, like the chalk downland snail *Vertigo pygmaea*, are common in open grassed areas such as lawns and meadows, whereas other snails like the *Planorbis leucostoma* frequently inhabit ponds and ditches. The courtyard soils at Chedworth Roman villa are dominated by the shells of *Pupilla* and *Vallonia* species of snail, which inhabit a dry, open and short-turfed grass environment, and intermediate species such as *Aegopinella*, *Discus* and *Trichia*, which thrive in damp shaded habitats. They suggest that the courtyard may well have been a lawned area, possibly with hedges, trees and the accompanying ground litter. The soil deposits in the Roman courtyard garden at Richebourg in France (see p. 90) contained charcoal, pottery sherds, pulverized mortar and mollusc shells, but the abundance of mollusc shells in these deposits may indicate an attempt to enrich the soil with calcium.[26] Virgil, in his first-century-BC treatise *Georgics* 2.346, recommended covering young plantings with topsoil or a thick layer of rough shells, and Palladius, in the fifth century AD, referred to the planting of certain types of trees on a 'bed of shells' (*Opus Agriculturae* 10.14). In the planting compartments on the garden terrace at the Villa of Livia at Prima Porta, 'cannibal' snails (*Rumina decollata*) may have been introduced to feed on land snails in the garden as a sort of pest control.[27] Pliny recommended seaweed manure for some vegetables (*Natural History* 19.41.143), and molluscs from marine environments in excavated soils could easily have been gathered up with the seaweed. Shells of molluscs associated with seaweed and other beach deposits have been found in archaeological contexts at Culver Street in Roman Colchester, for example, suggesting that seaweed may have been brought to the site as horticultural manure.[28]

In this chapter, we have focused on looking at the physical remains of plants and their scientific analysis. Such an analysis gives little insight into the appreciation that people might have had for their gardens and the natural environment. For an evocative appraisal of fruitful and flowering gardens and a deeper understanding of how people of the past interacted with nature, literature of the ancient world is invaluable. The next chapter looks at how gardens were represented in selected literary sources.

CHAPTER EIGHT

Gardens in ancient poetry

In various cultures of the ancient world, gardens and groves were praised in poetry. Of course the poems that survive are not meant to be actual historical records, yet they evoke what gardens might have been like. The range of vegetation named, the descriptions of the planted areas, and the kinds of activities that could take place in them reveal varying attitudes towards the garden. Some illustrate a particularly intimate relationship between man and the natural environment.

A common feature of Egyptian love poetry is the use of 'talking' trees that invite lovers to hold their romantic trysts in the orchard or garden. The combination of the shade of the branches, the colourful fruits and scented flowers creates a romantic setting. The types of tree encountered in the so-called *Turin Love Songs* are those most familiar in archaeology and art.[1]

Poem 1
The little sycomore,
which she planted with her hand,
sends forth its words to speak.

The flowers of its stalks
are like an inundation of honey;
beautiful it is, and its branches shine
more verdant than the grass.

It is laden with the ripeness of notched figs,
redder than carnelian,
like turquoise its leaves,
like glass its bark.

Its wood is like the colour of green feldspar,
its sap like the besbes opiate;
it brings near whoever is not under it,
for its shade cools the breeze.

It sends a message by the hand of a girl,
the gardener's daughter;
it makes her hurry to the lady love:
come, spend a minute among the maidens.

The country celebrates its day.
Below me is an arbour and a hideaway;
my gardeners are joyful
like children at the sight of you.

Poem 2
The pomegranate says:
like her teeth my seeds,
like her breasts my fruit,
foremost am I of the orchard
since in every season I'm around.

The sister and brother make holiday,
swaying beneath my branches;
high on grape wine and pomegranate wine are they,
and rubbed with Moringa pine oils . . .

All, all pass away,
except for me, from the fields.
Twelve months I spend
within the park waiting.

Where drops a flower,
another bud within me springs.
Of the orchard foremost am I, . . .
but only as a second you regard me.

Trees not only talk to each other, they also argue with each other on the topic of their usefulness in an Akkadian fable:[2]

> The king planted
> a palm in his palace.
> With it he planted a . . . tamarisk.
> In the shade of the tamarisk a dinner
> was given and in the shade of the palm . . .
> Each other's worth they insulted;
> the tamarisk and the date palm had a dispute.
> The tamarisk spoke thusly, 'I . . . greatly.
> Of the date palm is so wonderful . . . '.
> 'You, tamarisk, are a useless tree.
> What are your branches?
> Only wood without any fruit at all! . . .
> The gardener speaks well of me,
> of use to both slave and official.
> My fruit makes the infant grow,
> grown men also eat my fruit.'

Both trees and flowers speak to an unhappy lover in Persian poetry written after the Arab conquest by Farrukkhī:[3]

> My beloved left me, and I went to the garden, gloomy and lovesick because of it . . . 'If your darling has left you', said the violet, 'then pluck me and keep me in memory of her locks.' What did the narcissus say? It said, 'You are far from your lover's eyes, so let my eyes take your grief at not seeing hers.' I went on crying so much that I heard moaning and wailing from the cypresses. They confided in me and said, 'O that your heart could find peace with us, for your beloved was flourishing, and so are we. She was tall, and we are a hundred times taller.' 'You are tall and you are flourishing,' I said to the cypresses, 'but what use are you to me when it comes time for kissing?'
>
> (*Divan* 158)

Trees figure again in the so-called 'Parable of the Two Trees' in the ancient Egyptian *Instruction of Amenemope*. Here human nature is compared to the trees grown in different environments:[4]

> The hot-headed man in the temple
> Is like a tree grown in a garden;
> Suddenly it bears fruit.
> It reaches its end in the carpentry shop;
> It is floated away far from its place,
> Or fire is its funeral pyre.
>
> The truly temperate man sets himself apart,
> He is like a tree grown in a sunlit field,
> But it flourishes, it doubles its yield,
> It stands before its owner;
> Its fruit is something sweet, its shade is pleasant,
> And it reaches its end in a garden.

In Old Testament texts, human happiness and fecundity within the bonds of the family can be clothed in a metaphor of fruitful vines and fruit trees:

> For thou shalt eat the labour of thine hands: happy shalt thou be, and
> it shall be well with thee.
> Thy wife shall be as a fruitful vine by the sides of thine house;
> thy children like olive plants round about thy table.
>
> (Psalms 128.2–3)

In a rare celebration of physical intimacy in the Old Testament, the Song of Solomon makes frequent references to fruitful trees and nourishing waters in orchards and gardens. This natural environment is used to create a romantic setting:

> I am the rose of Sharon, and the lily of the valleys.
> As the lily among thorns, so is my love among the daughters.
> As the apple tree among the trees of the wood, so is my beloved
> among the sons.

I sat down under his shadow with great delight, and his fruit was
sweet to my taste.
He brought me to the banqueting house, and his banner over me
was love.
Stay me with flagons, comfort me with apples: for I am sick
of love.
His left hand is under my head, and his right hand doth
embrace me.

Awake, oh north wind, and come, thou south; blow upon my garden,
that the spices thereof may flow out. Let my beloved into his
garden, and eat his pleasant fruits.

My beloved is gone down into his garden, to the beds of spices,
to feed in the gardens, and to gather lilies . . .
I went down into the garden of nuts to see the fruits of the
valley, and to see whether the vine flourished, and the
pomegranates budded.

(Song of Solomon 2.1–6, 4.16, 6.2–3, 6.11)

Love poetry, delivered in song, is preserved in a Sumerian text, and the allegory of the garden and the human body is quite explicit:[5]

He has sprouted, he has burgeoned, he is lettuce
planted by the water,
My well-stocked garden of the . . . plain, my favoured
of the womb,
My grain luxuriant in its furrow – he is lettuce
planted by the water,
My apple tree which bears fruit up to its top –
he is lettuce planted by the water.

Greek poetry rarely conveys any sense of appreciation of, or pleasure in, the garden other than for its utilitarian value, but there are occasional fragments of poems in which the pleasurable environment of a grove or wooded area is mentioned. Generally, these refer to natural settings within sacred contexts, and there is no hint of any of the erotic pleasures alluded to in Egyptian or Hebrew poems. The following excerpts are taken from Sappho and Euripides:

> Hither to me from Crete, to this holy temple, where
> is your pleasant grove of apple trees, and altars
> fragrant with smoke of frankincense;
> Therein cold water babbles through apple branches, and
> the place is all shadowy with roses, and from the quivering
> leaves comes slumber down . . .
>
> (Sappho: fragment 2)

> Come, oh newly sprouted, oh
> Beautiful, serviceable shoot of laurel,
> Who sweeps the altar of Phoibos,
> Below the temple
> From the undying gardens
> Where the sacred spring waters drench,
> Spouting forth a continuing,
> Flowing stream,
> The holy myrtle foliage.
>
> (Euripides, *Ion*: lines 112–20)

The simple pleasures of nature are occasionally celebrated in Roman poetry. The Augustan poet Horace was one of the leading champions of the simple life, which he claimed to enjoy on his humble country estate. The following are taken from Horace's *Odes* and *Satires*:

Leave costly wreaths for lordly brows:
Of myrtle let my chaplet be;
Seek not for autumn's lingering rose;
Twine but the myrtle, boy, for me.

For all that blooms there's naught so fit
For thee, my boy, that pours't the wine;
For me, that quaff it as I sit,
O'erarched by this embowering vine.

(*Odes* 1.38)

This is what I prayed for! – a piece of land not so very large,
Where there would be a garden, and near the house a spring
Of ever-flowing water, and up above these a bit of woodland.
More and better than this have the gods done for me. I am content.

(*Satires* 2.6.1–4)

The Spanish-born Roman poet Martial expresses much the same sentiment about his country villa and the landscape in which it is set, and he prefers it to the gardens and orchards at the palace of the Phaeacian King Alkinoos praised by Homer (*Epigrams* 31).

This grove, these founts, this matted shade of arching vine, this conduit of refreshing water, and the meadows, and the beds of rose that will not yield to twice-bearing Paestum, and the pot-herb in January green, nor seared by frost; and the tame eel that swims in its shut tank, and the white dove-cote that harbours birds as white – these are my lady's gifts: to me returned after seven lustres has Marcella given this house and tiny realm. If Nausicaa were to yield me her sire's gardens, I could say to Alcinous 'I prefer my own.'

The female body and its desirability are likened by the Roman poet Catullus (*Carmina* 62) to a beautiful and untouched flower in a garden:

> Like a flower in a garden fenced by palisades,
> Safe from the herd's hooves and the ploughshare's bruises,
> Which suns mature, showers foster, winds caress,
> And boys and girls are eager to possess;
> Yet when sharp nails have nipped it and it fades,
> It's coveted by neither lads nor maids –
> Such is a girl: as long as she's untouched,
> Her family cherishes her; but once she loses
> Her fresh bloom and her body becomes smudged,
> Boys will reject her and girls like her less.

In medieval Persian poetry, garden imagery is often used to describe the physical appearance of girls and women. Fakhr al-Din Gurgani (*Vis u Ramin*) employs the image of paradise and garden plants to describe the girl Vis:[6]

> Sometimes it seemed that she were the garden of Paradise with fresh lilies growing in it. Her hair is like violets and her eyes like narcissus; her cheek is like the jonquil, and her face like the tulip. At other times it seemed as if she were a garden in autumn, with the fruits of October in it. Her black locks were like ripe grapes, her chin like an apple, and her breasts like pomegranates.

CHAPTER NINE

Gardens and paradises

Gardens have always been of fundamental importance to human existence in any culture of the Mediterranean, Near East and Europe. On one level, their cultivation and irrigation guaranteed the provision of fruit, herbs and vegetables for the inhabitants of farms, towns and villages. When cultivated on a larger scale, garden produce contributed to the economy.

Much hard work was involved in the care of gardens and plantings. In the ancient world, the soil and climatic conditions generally allowed vegetation to be cultivated and to flourish only where a natural source of water existed, and not until the Roman period did this change significantly with the extensive provision of water from aqueducts built over long distances. Divine protection of life, and therefore of vegetation and fertility, was celebrated in architecture, texts and art, and the temples and sanctuaries of many gods were provided with gardens and groves. But gardens, particularly the private ornamental gardens and parks filled with vegetation, as well as water features, sculpture and pavilions, were also symbols of prestige and affluence. The larger the available wealth, the larger the garden tended to be. In order to maintain the gardens and groves of the wealthy, skilled gardeners and landscape designers were employed, and in the Roman period they travelled as widely in the empire as the exotic plants and trees that were exported from the Near East to Italy and subsequently to the newly conquered lands in the West. For Assyrian and Egyptian kings and Roman aristocrats alike, the possession of even individual rare species of plants and trees could express status and privileged access.

The gardens under discussion here related to their surroundings – be it a busy urban neighbourhood or the hot and dusty steppes – but they were

Detail of fig. 90 (p. 127)
Humay approaches the palace of his beloved in a miniature from Baghdad, c. AD *1396.*

چو مه را ملک بر لب بام دید

89 Trees around a city are felled by an attacking army on this Assyrian relief from Nimrud, c. 728 BC.

also separated from them, most often by a boundary wall, fence or hedge. The enclosed precincts and groves of ancient Greece, the *temene*, were, taking the word literally, 'cut out of the landscape' (τέμνειν). The Persian word *pairidaeza* (paradise) is derived from *pairi* (around) and *daeza* (wall). In all civilizations in antiquity, gardens were enclosed spaces. The gardens of Egypt and Rome, when laid out in a courtyard, were bounded not by walls, but by ranges of buildings. Within this enclosed space, the garden was internally subdivided by plantings and other features. Rows of trees as linear features, hedges arranged in various designs, planting beds surrounding pools and artificial lakes, as well as espaliered trees and vines, created spatial volumes and defined visual relationships within the garden and between the garden and the surrounding architecture. The axial arrangement of planted trees, leading to Egyptian mortuary temples, flanking the paths in Persian royal *paradeisoi* or bisecting Roman villa parks, reflected and enhanced the architectural layout. They shaped the space and led those approaching through that space. Such garden designs posed a tangible contrast to untamed nature, although some untamed nature, in the form of mountains, rivers or the sea, was often drawn into the architectural setting by the deliberate siting of the buildings and their gardens and by the use of vistas.

Since gardens were frequently overt symbols of power and status, attacking armies would deliberately destroy groves and gardens, particularly in the Near East and Egypt (fig. 89). The Egyptian attack on the Hyksos capital of Avaris in the Nile delta involved the felling of trees at the palace.[1] The Egyptian pharaoh Thutmosis III in 1458 BC claimed in several campaigns to have felled all the 'pleasant trees' and gardens of his enemy in Syria, particularly those in the land of Naharin, so that the land became a place 'upon which there are no trees'.[2] The Assyrian King Shalmaneser III (859–824 BC) boasted that he destroyed the gardens of his foe, Hazael, at Damascus.[3] Ashurbanipal's army destroyed the sacred groves of Elamite Susa in 639 BC, and recorded this destruction: 'Their secret groves, into which no stranger ever penetrates, whose border he never oversteps – into these my soldiers entered, saw their mysteries, and set them on fire.'[4]

Cutting down fruit-bearing trees is condemned in the Book of Deuteronomy (20.29–30), but this did not pertain to 'trees which thou knowest that they be not trees for meat, thou shalt destroy and cut them down'; nevertheless, the Israelites destroyed all the trees of the Moabites (Second Book of Kings 3.25), including those that bore fruit. The first action taken by the Phoenician rebels against the Persian satrap of Artaxerxes in Sidon in 351–350 BC was the destruction of the royal park (Diodorus Siculus 16.41.5). In the late fifth century, Cyrus the Younger destroyed the palace and *paradeisos* of the Syrian ruler Belesys (Xenophon, *Anabasis* 1.4.11), and Pharnabazus, the Persian satrap of Daskyleion in Phrygia, lamented that his fine buildings and *paradeisoi* were all 'cut down and burned down' (Xenophon, *Hellenica* 4.1.33). The attacking army of Philip V caused widespread damage to the trees in the Nikephorion at Pergamon in 201 BC (Polybios 16.1.6). The Roman general Sulla, when besieging Athens in 86 BC and in need of timber, cut down the pride of this venerable city of learning, the shady groves of the shrines and gymnasia in the suburbs of Academy and Lykeion (Plutarch, *Sulla* 12.3). These few examples illustrate that groves and gardens, either connected with the royal palace or with the gods of a city, must have been viewed as emblematic of earthly authority and heavenly power. The destruction of them struck a blow to this power and physically and symbolically reduced it to ashes.

The fruit-bearing garden as a symbol of fecundity, pleasure and the promise of a better life figures particularly prominently in Christian beliefs. In this faith, such a garden is equated with paradise. The Christian image of paradise can be traced back to the Persian royal park (*pairidaeza*), later translated by the Greeks, possibly in the late fifth century BC, as *paradeisos*.[5] These Persian earthly paradises featured groves of trees, a river or other source of irrigation and wild animals. The translators of the Hebrew texts into a Greek compilation, the *Septuagint*, used the term *paradeisos* for the Hebrew *Gan Eden*, the garden of Eden, which loosely meant 'land of delight'.[6] In Hebrew texts such as the Song of Solomon, possibly recorded around 400 BC, the royal garden of Solomon planted with fruit trees is referred to with a Hebrew loanword as a *pardes*; in the Greek texts the royal garden of David in Jerusalem is a *paradeisos* (Josephus, *Jewish Antiquities* 7.347, 9.225). Solomon and David's *paradeisoi* and the contemporary royal *paradeisoi* of eastern monarchs may have been the garden paradises that the translators of the *Septuagint* envisaged when they translated the texts in Alexandria in the early third century BC. These were royal parks planted with trees, well-watered and suitable for contemplative strolls. God's gardener, Adam, 'heard the voice of the Lord God walking in the garden in the cool of the day', before he and Eve were banished from Eden for having eaten the fruit of the tree of knowledge (Genesis 3.8) (see frontispiece).

This story of paradise, the garden of God, describes the transformation from the golden age of man to the 'real' world in which man had been forced to forfeit a life of simple and eternal perfection for one of toil and strife. But paradise was not necessarily lost forever in the Christian faith, and, indeed, it could be regained, according to the teachings of the early Church fathers, when entering into the Church through baptism (Tertullian, *Adversus Marcionem* 2.4).[7] According to Cyril of Jerusalem, 'there is opened to thee the paradise of God, which He planted towards the East', where catechumens receive baptism (Cyril, *Procatechesis* 15.16). The Koran also contains frequent references to the image of paradise (*janna*) as a garden with flowing rivers and fountains, fruit trees and pavilions, and the garden of paradise awaits the faithful on the Day of Judgement, welcoming those who have believed in Allah and lived according to the teachings of Islam. Sometimes the Moslem paradise is referred to as the gardens of Eden.

Earthly gardens in both the Christian and Islamic faith were often perceived as reflections of paradise. The Byzantine emperor's palace garden was clearly equated with this heavenly paradise, as we can see in John Geometres' description of the Aretai palace near Constantinople in the second half of the tenth century AD. He asks, 'Who transferred the site of Eden here?', and he speaks of four springs flowing from the old Eden to water the new Eden.[8] Later, in medieval Europe, earthly paradise was presented as an image of the Church, and the monastery could be equated symbolically with paradise.[9] 'The

90 Humay approaches the palace of his beloved. The palace, with its observation platform, is set in a paradisiacal garden of fruit trees and flowers in this miniature from Baghdad, c. AD *1396.*

monastery bears the image of paradise, and an even more secure paradise than Eden. In this lovely garden the source of water is the baptismal font, the tree of life in this paradise is the body of the Lord. The various kinds of fruit trees are the different books of the Bible' (Honorius Augustodunensis, *Gemma anima* 1.149). In the Islamic world, royal gardens were often perceived as earthly versions of the garden at the beginning of time, some of them even excelling paradise (fig. 90).[10] The historian Khwandamir, for example, praised the floating Mughal garden of Humayun (reigned 1530–56), claiming that 'even the garden of paradise is not its equal'.[11]

The ideal garden, sustained by divine power, is not unique to Judaism, Christianity or Islam, nor were these the first religions to associate gardens with paradise. Sumerian mythology placed paradise in a divine and immortal garden as early as the third millennium BC.[12] In ancient Egypt, the tomb paintings of the deceased in a sumptuous garden being fed by Hathor bear witness to the belief in a better life after death, and in Greek and Roman religions, mythical fruit-bearing gardens existed as Elysium or the garden of the Hesperides. In these older religions, however, man had not been banished from a mythical, original and primeval garden, only to regain his position in it through good deeds and faith. Instead, the garden was a place in which one simply wished to spend eternity, hoping to enjoy the pleasures once experienced in an earthly paradise.

The gardens of the ancient world lived on in various forms and were sources of inspiration for other cultures in the centuries after the end of antiquity. The Arabs, conquering Persia in AD 651, encountered the Sassanian palace gardens and parks which, to them, must have appeared truly paradisiacal in contrast to the vast stretches of desert to which they were accustomed. They absorbed the Persian garden tradition, introducing variations of such gardens at their earliest capitals at Baghdad and Samarra in the eighth and ninth centuries. Under many Islamic dynasties thereafter the quadripartite royal gardens and the zoological parks of ancient Persia were copied and modified repeatedly from Egypt to Sicily and Spain (see Chapter 3).[13] Following the conquest of Persia in the late fourteenth century, Timur of Samarkand created luxurious Persian-inspired palace gardens – ten in total and all with Persian names – which deeply impressed visiting ambassadors from the West, particularly the Spanish ambassador, Ruy Gonzales de Clavijo. His account of these gardens in 1404 reveals the familiar features known in Persian gardens as early as the sixth century BC: enclosure walls, towers, belvederes, pavilions, fountains, pools and watercourses, paths, trees, vines and exotic animals.[14]

These same features appear in the extensive Topkapı palace of the Ottoman sultans in Istanbul in the fifteenth and sixteenth centuries, where a vast array of scented flowers was a particular source of pleasure and where deer, foxes, hares, sheep, goats, wild boar, bears, lions and various birds were kept and occasionally hunted.[15] The first Mughal emperor, Babur, left descriptions of his splendid four-fold garden south of Kabul, the Bagh-i Wafa, or Garden of Fidelity, planted in 1508 with oranges, citrons and pomegranates, in his autobiography, the *Babur-Nama* (fig. 91).[16] In all these gardens, the colour and fragrance of the vegetation and the abundance of flowing water are objects of particular admiration. Babur's successors in Mughal India in the sixteenth and seventeenth centuries were also prolific builders of gardens associated with palaces and dynastic tombs. A favourite design was the garden of quadripartite form, the *chahar bagh*, perhaps most perfectly and simply achieved at the monumental tomb complex of Mumtaz Mahal at Agra, known as the Taj Majal.[17]

Rome's former barbarian enemies in the West, the Vandals, revelled in the baths, theatres and the thrills of the hippodromes they inherited from the Romans after their acquisition of North Africa in the sixth century AD. Furthermore, they lived in well-watered *paradeisoi* filled with trees in which they held sumptuous banquets in the Roman tradition (Procopius, *Vandal Wars* 4.6.9). Roman gardening traditions also continued in the Byzantine world of the eastern Mediterranean. Gardens and parks, with their trees, flowers, fountains and pavilions, were integral parts of Byzantine palaces and villas, particularly in the capital of Constantinople.[18] Their appearance may be reflected in the wall mosaics in the Great Mosque in Damascus which was built around AD 715 by Caliph Al Walid, probably employing Byzantine mosaicists (fig. 92).

Both the Great Palace of the fourth century and the Blachernai Palace of the eleventh century in Constantinople were built in the tradition of earlier imperial palaces in the Roman West. Nero, for example, in AD 64 had designed his palace in Rome, the *Domus Aurea*, to include landscape gardens of

خوب می نماید خیلی باغ خوبی طرح شده و در طرف جنوب
کوه سفید ننگنهار واقع شده در میان ننگنهار و بنگش واسطه

درختهای انار همه سبز است گرد اگرد حوض تمام سبزه زار است
جای عمده باغ همین است در وقت زرد شدن نارنجها بسیار

vineyards, pastures, woodlands and animal parks and a lake surrounded by buildings made to resemble cities of the Roman Empire (Suetonius, *Nero* 31), and Hadrian in the second century AD constructed a vast palace at Tivoli in the midst of gardens, pools and pavilions, some of them intended to resemble famous landmarks in conquered Greece and Egypt. This microcosm of universal rule was also expressed in the Byzantine imperial residence where pavilions and palaces were built amidst gardens in various styles, including a Persian design, to symbolize an empire united under a sole monarch. It is no coincidence that the Ottoman sultan Mehmed II, who succeeded the Byzantine emperors, designed his New Palace, the Topkapı, in 1459 in Constantinople to include garden pavilions representing the Greek, Persian and Ottoman kingdoms united under his rule.[19]

In Byzantium, practical Roman agricultural literature continued to be compiled, and scenes of agricultural and horticultural labour were produced in illuminated manuscripts. The most famous of all Roman herbals was that written and illustrated by Dioscorides of Cilicia in the first century AD. This herbal, entitled *De Materia Medica*, and other illustrated manuals written after the first century, formed the basis for the compilation of plants commissioned in AD 512 as a gift for the Byzantine aristocrat Iuliana Anicia.[20] *De Materia Medica* was copied throughout the Byzantine period, and one of these copies was given as a gift by the Byzantine Emperor Romanos II in AD 948 to Caliph Abd al-Rahman III in Cordoba in Andalusia. Under the patronage of Romanos' father, Constantine VII Porphyrogenitos, an important contemporary treatise on agriculture and gardening, the *Geoponika*, was compiled. Roman illustrated manuscripts were equally influential in the Islamic world, and Dioscorides was translated into Arabic in about 854 and into Syriac in the thirteenth century. Surviving Arabic treatises on plants and their medicinal and practical

91 (Opposite) *The Mughal Emperor Babur oversees work conducted in the Garden of Fidelity (Bagh-i Wafa) in Kabul, in a miniature, c. AD 1595. The garden is subdivided by water channels and planted with oranges, pomegranates and other trees.*

92 (Above) *Detail of the wall mosaics in the Great Mosque in Damascus in Syria, c. AD 715, depicting gardens and villas.*

uses contain some excellent copies of Roman and Byzantine plant illustrations (fig. 93).

93 A grapevine in an illustrated Islamic herbal from northern Iraq or Syria, c. AD *1229.*

Also in the medieval West, Graeco-Roman agricultural texts and herbals were adopted and copied in monastic circles. In the sixth century AD, Dioscorides' *De Materia Medica* was translated from Greek into Latin in Italy, and it became widely disseminated in western Europe in the centuries that followed. Another herbal written by Apuleius in the fourth century AD was also translated into Latin, and various versions of this manuscript, such as an early twelfth-century herbal made at Bury St Edmunds, were prepared in monastic centres in the West throughout the Middle Ages (fig. 94).

Roman peristyle courtyards appear to have been the architectural source of inspiration for the cloisters of early medieval monasteries, and, like the Roman prototypes, the open area in the cloister was planted, according to Hugh of Fouilloy (*De claustro animae* 34–5) and other documentary evidence, with grass or ivy and a tree, symbolizing the tree of life, in the centre. But even at these monastic cultural centres, as we can see at the monastery of St Gall in Switzerland in the early ninth century, gardens were primarily utilitarian and designed to supply the inhabitants with herbs, medicines, fruit and vegetables. The surviving plan of this monastery shows a kitchen garden with vegetable beds, a garden with beds for medicinal herbs and an orchard which doubled as a cemetery.[21] Many of the plants grown in medieval monasteries had been introduced to Europe by the Romans, and Roman herbals describing the healing properties of these plants were kept as reference works in monastic libraries.

By the thirteenth century, secular gardens associated with courtly culture began to incorporate not only beds for herbs and fruit trees, but also lawns and turf banks for pleasurable sojourns in the garden.[22] The treatise on gardening by Pietro Crescenzi of Bologna (*Opus Ruralium Commodorum*), written around 1300, contains descriptions of contemporary gardens ranging from the small herb garden, through gardens of medium size with turf seats and hedges of roses, pomegranate, plum and quince trees, to the large parks of kings and noblemen. Not until the Renaissance period, however, did Western garden designers combine the written descriptions of Roman villa gardens with the recently gained knowledge of ancient sculptures unearthed at various sites, to consciously create stately gardens in the spirit of the Classical past.[23] The physical shaping and cultural manipulation of nature reached a climax in the Baroque period in the ornamental design of parks and gardens inspired by Roman topiary and formal landscape design.

Whatever the evidence – be it preserved in the soil, in the visual arts, in prose or in song – the gardens of the ancient world offer a remarkable insight into the natural and created environments of the peoples of the past. Even if the picture that we have of ancient horticulture is imperfect, due mainly to the perishable nature of the evidence, it is clear that gardens and the cultivation of vegetation played a vital role in the daily lives of people of the ancient world. Our fascination in earthly paradises is kept alive as new traces and remains of ancient gardens are discovered, especially through archaeological exploration.

94 A thistle plant depicted in a medieval illustrated herbal from Bury St Edmunds, c. AD 1120.

Notes

Chapter 1

1 Carroll-Spillecke (1994); Hellenkemper Salies (1994).
2 Jashemski (1979).
3 Carroll-Spillecke (1989), 25f., 36.
4 Carroll-Spillecke (1989), 58.
5 Haller and Andrae (1955).
6 Thompson (1937).
7 Jashemski (1995), 559–72.
8 Murphy and Scaife (1991), 88.
9 Murphy and Scaife (1991), 92.
10 Pailler (1987).

Chapter 2

1 Vercoutter (1965).
2 Kemp (1987), 50–2, figs 4.5–6; Kemp (1985), 29–35, fig. 3.2. On similar garden beds at Amarna see Griffith (1924), 303, pl. 34.2.
3 Zohary and Hopf (1993), 237.
4 Borchardt and Ricke (1980); Kemp (1989), 294–8.
5 Lichtheim (1975), 24.
6 Dziobek (1992), 61–2.
7 Bietak (1996), 22, pl. 8.
8 Peet and Woolley (1923), 111–19; Badaway (1956).
9 Simpson (1973), 308f.
10 Margueron (1992), 74–9; Dalley (1993).
11 Leach (1982).
12 Carroll-Spillecke (1992a), 84f.
13 Carroll-Spillecke (1989).
14 Carroll-Spillecke (1989), 67–9.
15 Carroll-Spillecke (1989), 63–5.
16 Brown (1980), 64–9, figs 81–3, 86, 89–91.
17 Jashemski (1987), 64–70, fig. 28.
18 Jashemski (1981), 32–4, fig. 9.
19 Farrar (1998), 19f., and 64–84 on pools.
20 Jashemski (1979), 201–65.
21 Jashemski (1987); De Caro (1987).
22 Klynne and Liljenstolpe (2000), 224f.
23 De Alarcâo and Etienne (1987), 69–71, figs 1, 11–12.
24 Cunliffe (1971), 120–40.
25 Ebnöther (1995).
26 Branigan (1971), 85f.; Thomas (1964), 201–10; Crummy (1984), 138–41, fig. 129.

Chapter 3

1 Gale *et al.* (2000), 335–52; Murray (2000), 616–27.
2 Hugonot (1992), 39.
3 Schulz and Sourouzian (1998), 160, fig. 25.
4 Breasted Vol. 4 (1906/7), 333. On the import of myrrh trees see Dixon (1969) and Saleh (1972).
5 Luckenbill Vol. 1 (1989), 87; Grayson Vol. 2 (1972), 33.
6 Pritchard (1969), 558f.; Grayson Vol. 2 (1972), 174.
7 Luckenbill Vol. 2 (1989), 42.
8 Wendrich (1990).
9 Moynihan (1979), 14–27; Carroll-Spillecke (1989), 39, 58f.
10 Pinder-Wilson (1976), Moynihan (1979), 28–37; Kawami (1992), 93–8.
11 Stronach (1989); Stronach (1994).
12 Carroll-Spillecke (1989), 28–31.
13 Zanker (1979).
14 Neudecker (1988), 105–14, 147–57; Dillon (2000), 22–30.
15 Brown (1980), 24, fig. 21.
16 Rostovtzeff (1922).
17 Gleason (1994).
18 http://www.comune.roma.it/cultura/italiano/musei_spazi_espositivi/musei/museo_fori/aree/fs2pace.htm; Lloyd (1982).
19 Richardson (1992), 195–204.
20 Tomei (1992); Villedieu (2001), 60–76.

Chapter 4

1 Hugonot (1992), 37f., fig. 13; Wilkinson (1998), 74–87; Schulz and Sourouzian (1998), 184–8.
2 Hugonot (1992), 36.
3 Kemp (1987), 59f., figs 5.3–4; Kemp (1985), 47–9, figs 4.6–7.
4 Hugonot (1992), 22f., fig. 7.
5 Margueron (1992), 61.
6 Kawami (1992), 82.
7 Bietak (1996), 36, fig. 31, pl. 14.
8 Karageorghis and Carroll-Spillecke (1992).
9 Soren (1987), 42–4; Carroll-Spillecke (1992a), 90f., fig. 15.
10 Miller (1977), 11; Miller (1978), 65.
11 Roebuck (1951), 41.
12 Camp (1986), 40–2.
13 Kent (1948).
14 Lauter (1968/69).
15 Andreae (1957), 165, fig. 21.
16 Jashemski (1979), 157f., fig. 244.
17 Carroll and Godden (2000).
18 Jashemski (1995), 573, fig. 14.
19 Levi (1967), 155f., Seg. IX.

Chapter 5

1 Kampp-Seyfried (1998), 250f., figs 197–9.
2 Willeitner (1998), 314f., fig. 47.
3 Carroll-Spillecke (1989), 58.
4 Laum (1914); Fraser and Nicolas (1958).
5 Andronicos (1984), 97–118, figs 57–63.
6 Toynbee (1971), 94–100.
7 Toynbee (1971), 97f.
8 Secchi (1843).
9 Toynbee (1971), 144, pls 49–50.
10 Jashemski (1970/71).

Chapter 6

1 Wente (1990), 92f.
2 Murray (2000), 616.
3 Rostovtzeff (1922); Carroll-Spillecke (1989), 54–6, 58–60.
4 Margueron (1922), 58.
5 Parker (1941).
6 Jashemski and Ricotti (1992), 585–95, figs 9–15.
7 Carroll-Spillecke (1989), 72–5.
8 Carroll-Spillecke (1989), 72.
9 Jashemski (1974), 399–400, pl. 81.9.
10 Jashemski (1987), 73.
11 Barat and Morize (1999); Klynne and Liljenstolpe (2000), 221–4, figs 3–4.
12 Detsicas (1981).
13 Villedieu (2001), 84–98, figs 76–80.
14 Jashemski and Ricotti (1992), 580–5, figs 5–7.
15 Tomei (1992), 938–43; Villedieu (1992), 465–79; Lloyd (1982), 95–100.
16 Atallah (1966).
17 Thompson (1937).
18 Carroll-Spillecke (1989), 47f., fig. 21.
19 White (1977).

20 Carroll-Spillecke (1989), 75.
21 Jashemski (1979a), 406f., pls 58.5–6.
22 Zimmer (1982), 180–2; Horn (1987), 161, fig. 95.
23 White (1977), 27f.
24 Jashemski (1981), 32, fig. 9.
25 Ciarallo and Mariotti Lippi (1993).
26 Carroll-Spillecke (1989), 56.

Chapter 7

1 Zohary and Hopf (1993), 141; Murray (2000), 614.
2 Moynihan (1979), 42; Ruggles (2000), 17f.
3 Dickson (1994); Murphy and Scaife (1991); Willcox (1977); Knörzer (1966).
4 Jashemski (1974), 401f.
5 Filtzinger, Planck and Cämmerer (1986), 428f.; Becker and Tegtmeier (1998).
6 Jacomet, Kučan, Ritter, Suter and Hagendorn (2002).
7 Hepper (1990), 9–10; Manniche (1989), 22–32.
8 Renfrew (1985).
9 Dickson (1994), 54.
10 Dickson (1994), 55.
11 Dickson (1994), 54.
12 Dickson (1994), 55.
13 Dickson (1994), 56–8; Murphy and Scaife (1991), 88.
14 Kučan (1992).
15 Murphy (1984), 40.
16 Bakels and Jacomet (2003).
17 Gabriel (1955).
18 *Il Giardino Dipinto.*
19 Stauffer (1991), 35–53.
20 Carroll-Spillecke (1993), 36, 145. On the written records of the monastic diet see Bond (2001).
21 Carroll-Spillecke (1989), 44.
22 Hoepfner and Schwandner (1986), 40, 156f.
23 Miller and Gleason (1994), 38f.; Zeepvat (1991), 55f.
24 Gracie and Price (1979), 13.
25 Murphy and Scaife (1991), 84.
26 Barat and Morize (1999), 223.
27 Klynne and Liljenstolpe (2000), 230.
28 Murphy (1992), 276f.

Chapter 8

1 Simpson (1973), 312, 314.
2 Pritchard (1969), 593.
3 Hanaway (1976), 55.
4 Simpson (1973), 146.
5 Pritchard (1969), 645.
6 Hanaway (1976), 53.

Chapter 9

1 Bietak (1996), 65.
2 Pritchard (1955), 240.
3 Pritchard (1955), 280.
4 Luckenbill Vol. 2 (1989), 310.
5 Bremmer (1999).
6 Noort (1999).
7 Benjamins (1999).
8 Maguire (1990).
9 Meyvaert (1986), 51f.
10 Macdougall and Ettinghausen (1976); Barrucand (2000), 490.
11 Wescoat (1986).
12 Moynihan (1979), 3.
13 Tabbaa (1992); Macdougall and Ettinghausen (1976); Wilber (1979); Ruggles (2000).
14 Wilber (1979), 23–37.
15 Necipoğlu (1991).
16 Moynihan (1979), 79–86.
17 Jellicoe (1976); Moynihan (1979), 96–147.
18 Littlewood (1992); Littlewood (1997).
19 Necipoğlu (1991), 211–12, 245–50.
20 Singer (1927); Gunther (1959).
21 Price (1982); Willerding (1992).
22 Hennebo (1987); Landsberg (1995).
23 Macdougall (1994).

Gardens to visit

The following is an overview of Roman gardens in Britain and on the continent that have been replanted and are worth a visit. Reconstructed medieval gardens worth visiting are listed in S. Landsberg, *The Medieval Garden*, London, British Museum Press, 1995, pp. 139–40.

BRITAIN

Fishbourne Roman Palace

This site near Chichester in southern England includes the consolidated remains of a palatial Roman house of the later first century AD. Outside the roofed remains of the house and its mosaic pavements is the excavated Roman garden, replanted with a lawn, clipped box hedges and espaliered trees, according to the planting patterns detected as bedding trenches and post-holes in the archaeological excavations in the 1960s. There is also an area set aside for an experimental garden in which plants of the Roman period are grown. A display in the small Museum of Roman Gardens on site deals with Roman horticulture, gardening tools and planting pots.

Chedworth Roman Villa

The partially excavated and consolidated remains of this villa near Cirencester in Gloucestershire are owned by the National Trust. Excavations from 2000 to 2002 in the upper courtyard of the villa have shown that this area was laid out as a lawned garden in the early fourth century AD. The grassed courtyard probably resembles the Roman garden, although there may have been garden features such as statues or a pool that have not survived. Surrounding the courtyard are roofed dining rooms and baths, many of them with mosaic floors.

Corinium Museum, Cirencester

A small area to one side of the museum has been laid out as a courtyard garden of a Roman house, complete with growing plants, water features, sculptures and wall paintings. The reconstruction is based on excavated evidence from many Roman gardens on the continent.

FRANCE

St-Romain-en-Gal

This site is situated on the left bank of the Rhône river, opposite the Roman town of Vienne. Excavations uncovered the remains of numerous houses and workshops, as well as paved Roman streets. The most impressive of these houses is the so-called Maison des Dieux Océans which has two courtyards laid out as gardens with pools and fountains. The gardens have been replanted, and the pools and fountains are fully functional. Next to the site is a large modern museum with excellent displays on the history and archaeology of the Roman settlement on both sides of the Rhône.

ITALY

Pompeii

Just about every courtyard in Pompeii was planted as a garden when Vesuvius erupted in AD 79, so it is difficult to recommend just a few. Some of them were replanted years ago, but they are often completely overgrown and neglected, or the houses have been closed to the public. The following, however, are accessible and worth a visit:

A reconstructed Roman garden at Corinium Museum, Cirencester.

Fountain in the form of a bronze statue of a boy carrying a goose in the garden of the House of the Vettii, Pompeii.

House of the Vettii (Casa dei Vettii)
One enters this house of the mid-first century AD through a rather dark atrium, beckoned by the light and the glimpse of a green garden at the back. Surrounded by columned walkways on all four sides, the garden is a profusion of flowering and creeping plants and shrubs that meander their way around the courtyard. The planting pattern is meant to reproduce that discovered in excavations of the house in the nineteenth century. Various garden sculptures in marble and bronze stand in the courtyard and at both ends of the garden, some of them functioning as fountains. Around the edges of the courtyard are the original lead pipes that once piped water into the house and irrigated the garden.

House of the Faun (Casa del Fauno)
One of the largest patrician houses in Pompeii, and named after a bronze sculpture of a dancing faun, this grand residence was remodelled and enlarged in the second and first centuries BC. It has two peristyle courtyards. The first courtyard beyond the atrium has been replanted with box hedges in geometric patterns. The larger, rear courtyard, which may have been a utilitarian garden or orchard, today has a grassed surface and a few large trees.

House of Loreius Tiburtinus (Casa di Loreio Tiburtino)
This house, located near the amphitheatre, is modest, but it once had a planted atrium, a tiny peristyle garden beyond the atrium and a very large garden at the back. Both peristyle gardens have been replanted, but the large garden is the most spectacular of the two. It has a T-shaped channel, the long arm of which cuts down the middle and runs from one end of the garden to the other. The monotony of the channel is broken by a stepped fountain, a pavilion and basin. On either side of the channel was a wooden arbour, which today has been replaced, and the edges of the garden were lined with trees.

House of Venus Marina (Casa di Venere)
The courtyard of this house is planted as a garden, although the garden design may not reflect the original layout. Of interest are the paintings on the back wall of the courtyard depicting statues in a luxurious garden enclosed within a fence.

House of Hercules (Giardino d'Ercole)
The large garden attached to this house was excavated in the 1970s. In the Roman period it was planted with flowers, roses and olive trees for the production of perfumed oils. The garden is attractively replanted, and it is worth visiting to see the masonry dining couches (*triclinium*) protected from the sun by a shady vine pergola.

House in Regio I.20.5
In the south-eastern part of Pompeii many houses had large vineyards and orchards attached to them. This modest house had a plot of land planted with vines for the purpose of making wine on the premises. Large wine vats sunk into the ground are still visible. The vines were replanted a few years ago to give the visitor an impression of the original appearance of a purely utilitarian garden.

Torre Annunziata (Oplontis)
Villa of Poppaea
One of the most beautiful luxury Roman villas of the Naples area, this perhaps belonged to the Emperor Nero's wife Poppaea. Of the many gardens excavated here in the 1970s, the loveliest is the large park-like garden at the back of the villa, replanted with oleanders, trees and box hedges, as it was in antiquity. It is best visited in May when the oleanders are a mass of pink flowers.

Boscoreale
Roman farm at Villa Regina
This modest farmhouse and the land around it was excavated in the 1970s and 1980s and has since been replanted with vines in almost exactly the same position as the original Roman vines. The vineyard comes right up to the farmhouse, and there are some impressive concrete casts of large trees that were buried when Vesuvius erupted in AD 79. Lying well below the modern ground level, it is also one of the best sites to visit to appreciate the amounts of ash and pumice that buried the farms and towns at the foot of the volcano. A museum next to the villa has an excellent display of artefacts from and information on the area.

PORTUGAL

Conimbriga
The Roman town of Conimbriga once had impressive Roman houses designed with mosaic floors and interior courtyard gardens, some of which have been excavated. The most interesting is the House of the Jets of Water, named after the still-functioning fountains piping water into a pool in the courtyard. Six ornamental masonry containers in the pool were planted in the Roman period with flowers and shrubs; they are still planted today. There is a site museum exhibiting excavated finds from the site.

Replanted peristyle garden in the House of the Faun, Pompeii, with a view towards the large garden beyond.

Replanted vineyard (House I.20.5), Pompeii, with large wine vats in the foreground.

Bibliography

Andreae, B., 1957, Archäologische Funde im Bereich von Rom 1949–56/57, *Archäologischer Anzeiger* 72, pp. 110–358.

Andronicos, M., 1984, *Vergina. The Royal Tombs*, Athens.

Atallah, W., 1966, *Adonis dans la littérature et l'art grecs*, Paris.

Badaway, A., 1956, 'Maru-Aten: Pleasure resort or temple?', *Journal of Egyptian Archaeology* 42, pp. 58–64.

Bakels, C., and Jacomet, S., 2003, 'Luxury foods in Roman period Central Europe', in M. van der Veen (ed.), *Luxury Foods* (World Archaeology 34/3).

Barat, Y., and Morize, D., 1999, 'Les pots d'horticulture dans le monde antique et les jardins de la villa Gallo-Romaine de Richebourg (Yvelines)', *Société française d'étude de la céramique antique en Gaule. Actes du Congrès de Fribourg*, Marseilles, pp. 213–35.

Barrucand, M., 2000, 'The Garden as a Reflection of Paradise', in Hattstein and Delius, pp. 490–3.

Becker, W.-D., and Tegtmeier, U., 1998, 'Datteln, Feigen, Mandeln, Nüsse und Südfrüchte aus dem römischen Xanten', *Archäologie im Rheinland 1997*, pp. 188–91.

Benjamins, H.S., 1999, 'Paradisiacal Life: The Story of Paradise in the Early Church', in Luttikhuizen, pp. 153–67.

Bietak, M., 1996, *Avaris. The Capital of the Hyksos*, London.

Bond, J., 2001, 'Production and Consumption of Food and Drink in the Medieval Monastery', in G. Keevill, M. Aston and T. Hall (eds), *Monastic Archaeology. Papers on the Study of Medieval Monasteries*, Oxford, pp. 54–87.

Borchard, L., and Ricke, H., 1980, *Die Wohnhäuser in Tell el-Amarna*, Berlin.

Branigan, K., 1971, *Latimer. Belgic, Roman, Dark Age and early modern farm*, Bristol.

Breasted, J.H., 1906–7, *Ancient Records of Egypt*, Vol. 4, Chicago.

Bremmer, J.N., 1999, 'Paradise: from Persia, via Greece, into the *Septuagint*', in Luttikhuizen, pp. 1–20.

Brown, A.E. (ed.), 1991, *Garden Archaeology*. Council for British Archaeology, Research Report 78, London.

Brown, F., 1980, *Cosa. The Making of a Roman Town*, Ann Arbor.

Camp, J.M., 1986, *The Athenian Agora. Excavations in the heart of Classical Athens*, London.

Carroll, M., 2001, *Romans, Celts and Germans. The German Provinces of Rome*, Stroud.

Carroll, M., and Godden, D., 2000, 'The Sanctuary of Apollo at Pompeii: Reconsidering Chronologies and Excavation History', *American Journal of Archaeology* 104, pp. 743–54.

Carroll-Spillecke, M., 1989, Κῆπος. *Der antike griechische Garten*, Munich.

Carroll-Spillecke, M. (ed.), 1992, *Der Garten von der Antike bis zum Mittelalter*, Mainz (reprinted 1995 and 1998).

Carroll-Spillecke, M., 1992a, 'The gardens of Greece from Homeric to Roman times', *Journal of Garden History* 12.2, pp. 84–101.

Carroll-Spillecke, M., 1993, *Die Untersuchungen im Hof der Neuen Universität in Heidelberg, Tiefgarage der Universitätsbibliothek*, Stuttgart.

Carroll-Spillecke, M., 1994, *Der Mahdia-Garten im Rheinischen Landesmuseum Bonn*, Cologne.

Ciarallo, A., and Mariotti Lippi, M., 1993, 'The garden of "Casa dei Casti Amanti" (Pompeii, Italy)', *Garden History* 21.2, pp. 110–15.

Connolly, P., and Dodge, H., 1998, *The Ancient City. Life in Classical Athens and Rome*, Oxford.

Crummy, P., 1984, *Excavations at Lion Walk, Balkerne Lane, and Middleborough, Colchester, Essex*. Colchester Archaeological Reports 3, Colchester.

Cunliffe, B., 1971, *Excavations at Fishbourne 1961–9*, London.

Dalley, S., 1993, 'Ancient Mesopotamian gardens and the identification of the hanging gardens of Babylon resolved', *Garden History* 21.1, pp. 1–13.

De Alarcâo, J., and Etienne, R., 1987, 'Les Jardins à Conimbriga (Portugal)', in Macdougall, pp. 67–80.

De Caro, S., 1987, 'The Sculptures of the Villa of Poppaea at Oplontis: A preliminary report', in Macdougall, pp. 77–134.

De Garis Davies, N., 1927, *Two Ramesside Tombs at Thebes*, New York.

Delumeau, J., 1995, *History of Paradise: The garden of Eden in myth and tradition*, Urbana.

Detsicas, A.P., 1981, 'A group of pottery from Eccles, Kent', in *Roman Pottery Research in Britain and Northwest Europe*. British Archaeological Reports, International Series 123, Oxford, pp. 441–5.

Dickson, C., 1994, 'Macroscopic Fossils of Garden Plants from British Roman and Medieval Deposits', in D. Moe, J.H. Dickson, P.M. Jorgensen (eds), *Garden History: Garden Plants, Species, Forms and Varieties from Pompeii to 1800*. European Symposium in Ravello, Rixensart, pp. 47–72.

Dillon, S., 2000, 'Subject selection and viewer reception of Greek portraits', *Journal of Roman Archaeology* 13, pp. 21–40.

Dix, B., 1994, 'Garden Archaeology at Kirby Hall and Hampton Court', *Current Archaeology* 12, No. 140, pp. 292–9.

Dixon, D.M., 1969, 'The transplantation of Punt incense trees in Egypt', *Journal of Egyptian Archaeology* 55, pp. 55–65.

Dziobek, E., 1992, *Das Grab des Ineni. Theben Nr. 81*, Mainz.

Ebnöther, C., 1995, *Der römische Gutshof in Dietikon*, Zürich.

Farrar, L., 1998, *Ancient Roman Gardens*, Stroud.

Filtzinger, P., Planck, D., and Cämmerer, B., 1986, *Die Römer in Baden-Württemberg*, Stuttgart.

Fraser, P.M., 1984, *Ptolemaic Alexandria*, Oxford.

Fraser, P.M., and Nicolas, B., 1958, 'The funerary garden of Mousa', *Journal of Roman Studies* 48, pp. 117–29.

Gabriel, M.M., 1955, *Livia's Garden Room at Prima Porta*, New York.

Gale, R., Gasson, P., Hepper, N., and Killen, G., 2000, 'Wood', in Nicholson and Shaw, pp. 334–71.

Ghaffar Shedid, A., and Seidel, M., 1991, *Das Grab des Nacht*, Mainz.

Il Giardino Dipinto nella casa del Bracciale d'Oro a Pompei, 1991, Exhibition Catalogue Centro Mostre di Firenze, Florence.

Gleason, K.L., 1994, 'Porticus Pompeiana: a new perspective on the first public part of ancient Rome', *Journal of Garden History* 14.1, pp. 13–27.

Gracie, H.S., and Price, E.G., 1979, 'Frocester Court Roman Villa, Second Report 1968–77: The Courtyard', *Bristol and Gloucestershire Archaeological Society Transactions* 97, pp. 9–64.

Grayson, A.K., 1972, *Assyrian Royal Inscriptions*, 2 Vols, Wiesbaden.

Griffith, F.L., 1924, 'Excavations at El-'Amarnah', *Journal of Egyptian Archaeology* 10, pp. 299–305.

Gunther, R.T., 1959, *The Greek Herbal of Dioscorides*, New York.

Haller, A., and Andrae, W., 1955, *Die Heiligtümer des Gottes Assur und der Sin-Samas-Tempel in Assur*, Berlin.

Hanaway, W.L., 1976, 'Paradise on Earth: The terrestrial Garden in Persian Literature', in Macdougall and Ettinghausen, pp. 41–68.

Hattstein, M., and Delius, P. (eds), 2000, *Islam. Art and Architecture*, Cologne.

Hellenkemper Salies, G., 1994, *Das Wrack. Der antike Schiffsfund von Mahdia*, Cologne.

Hennebo, D., 1987, *Gärten des Mittelalters*, Munich.

Hepper, F.N., 1990, *Pharaoh's Flowers. The Botanical Treasures of Tutankhamun*, Kew.

Hoepfner, W., and Schwandner, E.-L., 1986, *Haus und Stadt im klassischen Griechenland*, Munich.

Horn, H.G. (ed.), 1987, *Die Römer in Nordrhein-Westfalen*, Stuttgart.

Hugonot, J.-C., 1992, *Ägyptische Gärten*, in Carroll-Spillecke, pp. 9–44.

Hunt, J.D. (ed.), 1992, *Garden History: Issues, Approaches, Methods*. Dumbarton Oaks Colloquium on the History of Landscape Architecture XIII, Washington.

Jacomet, S., Kučan, D., Ritter, A., Suter, G., and Hagendorn, A., 2002, 'Pomegranates (*Punica granatum* L.) from early Roman contexts in Vindonissa (Switzerland)', in S. Jacomet, G. Jones and M. Charles (ed.), *Vegetation History and Archaeobotany*, Sheffield, pp. 79–92.

Jashemski, W.F., 1967, 'The Caupona of Euxinus at Pompeii', *Archaeology* 20/1, pp. 36–44.

Jashemski, W.F., 1970/71, 'Tomb gardens at Pompeii', *Classical Journal* 66/2, pp. 97–115.

Jashemski, W.F., 1974, 'The discovery of a market-garden orchard at Pompeii: The garden of the "House of the Ship Europa"', *American Journal of Archaeology* 78, pp. 391–404.

Jashemski, W.F., 1979, *The Gardens of Pompeii, Herculaneum and the Villas Destroyed by Vesuvius*, New Rochelle.

Jashemski, W.F., 1979a, 'The Garden of Hercules at Pompeii (II.viii.6). The discovery of a commercial flower garden', *American Journal of Archaeology* 83, pp. 404–11.

Jashemski, W.F., 1981, 'The Campanian Peristyle Garden', in Macdougall and Jashemski, pp. 31–48.

Jashemski, W.F., 1987, 'Recently excavated gardens and cultivated land of the villas at Boscoreale and Oplontis', in Macdougall, pp. 33–75.

Jashemski, W.F., 1992, 'The Contribution of Archaeology to the Study of Ancient Roman Gardens', in Hunt, pp. 5–30.

Jashemski, W.F., 1995, 'Roman Gardens in Tunisia: Preliminary Excavations in the House of Bacchus and Ariadne and in the East Temple at Thuburbo Maius', *American Journal of Archaeology* 99, pp. 559–76.

Jashemski, W.F., and Ricotti, E.S.P., 1992, 'Preliminary Excavations in the Gardens of Hadrian's Villa: The Canopus Area and the Piazza d'Oro', *American Journal of Archaeology* 96, pp. 579–97.

Jellicoe, S., 1976, 'The Development of the Mughal Garden', in MacDougall and Ettinghausen, pp. 107–29.

Kampp-Seyfried, F., 1998, 'Overcoming Death – The Private Tombs of Thebes', in Schulz and Seidel, pp. 249–63.

Karageorghis, V., 1995, *The Coroplastic Art of Ancient Cyprus IV. The Cypro-Archaic Period*, Nicosia.

Karageorghis, V., and Carroll-Spillecke, M., 1992, 'Die heiligen Haine und Gärten Zyperns', in Carroll-Spillecke, pp. 141–52.

Kawami, T.S., 1992, 'Antike persische Gärten', in Carroll-Spillecke, pp. 81–100.

Kemp, B.J., 1985, *Amarna Reports II*. Egypt Exploration Society, London.

Kemp, B.J., 1987, *Amarna Reports IV*. Egypt Exploration Society, London.

Kemp, B.J., 1989, *Ancient Egypt. Anatomy of a Civilization*, London.

Kent, J.H., 1948, 'The temple estates of Delos, Rheneia and Mykonos', *Hesperia* 17, pp. 243–338.

Klynne, A., and Liljenstolpe, P., 2000, 'Investigating the gardens of the Villa of Livia', *Journal of Roman Archaeology* 13, pp. 220–33.

Knörzer, K.-H., 1966, 'Über Funde römischer Importfrüchte in Novaesium', *Bonner Jahrbuch* 166, pp. 424–43.

Kučan, D., 1992, 'Die Pflanzenreste aus dem römischen Militärlager Oberaden', in J.S. Kühlborn (ed.), *Das Römerlager in Oberaden*, Vol. 3, Münster, pp. 237–66.

Landsberg, S., 1995, *The Medieval Garden*, London.

Laum, B., 1914, *Stiftungen in der griechischen und römischen Antike* II, Berlin.

Lauter, H., 1968/69, 'Ein Tempelgarten?', *Archäologischer Anzeiger*, pp. 626–31.

Leach, H., 1982, 'On the origins of kitchen gardening in the Ancient Near East', *Garden History* 10.1, pp. 1–16.

Levi, A., 1967, *Itineraria picta: Contributo allo studio della Tabula Peutingeriana*, Rome.

Lichtheim, M., 1975, *Ancient Egyptian Literature*, Vol. 1, Berkeley.

Littlewood, A.R., 1992, 'Gardens of Byzantium', *Journal of Garden History* 12.2, pp. 126–53.

Littlewood, A.R., 1997, 'Gardens of the Palaces', in H. Maguire (ed.), *Byzantine Court Culture from 829 to 1024*, Washington, pp. 13–38.

Lloyd, R.B., 1982, 'Three Monumental Gardens on the Marble Plan', *American Journal of Archaeology* 86, pp. 91–100.

Luckenbill, D.D., 1989, *Ancient Records of Assyria and Babylonia*, 2 Vols, London.

Luttikhuizen, G.P. (ed.), 1999, *Paradise Interpreted. Representations of Biblical Paradise in Judaism and Christianity*, Leiden.

Macdougall, E.B. (ed.), 1986, *Medieval Gardens*. Dumbarton Oaks Colloquium on the History of Landscape Architecture IX, Washington.

Macdougall, E.B. (ed.), 1987, *Ancient Roman Villa Gardens*. Dumbarton Oaks Colloquium on the History of Landscape Architecture X, Washington.

Macdougall, E.B., 1994, *Fountains, Statues and Flowers: Studies in Italian Gardens of the 16th and 17th centuries*, Washington.

Macdougall, E.B., and Ettinghausen, R. (eds), 1976, *The Islamic Garden*. Dumbarton Oaks Colloquium on the History of Landscape Architecture IV, Washington.

Macdougall, E.B., and Jashemski, W.F. (eds), 1981, *Ancient Roman Gardens*. Dumbarton Oaks Colloquium on the History of Landscape Architecture VII, Washington.

Maguire, H., 1986, *Earth and Ocean. The Terrestrial World in Early Byzantine Art*, University Park (Pennsylvania).

Maguire, H., 1990, 'A description of the Aretai palace and its garden', *Journal of Garden History* 10.4, pp. 209–13.

Manniche, L., 1989, *An Ancient Egyptian Herbal*, London.

Margueron, J.-C., 1992, 'Die Gärten im Vorderen Orient', in Carroll-Spillecke, pp. 45–80.

Meyvaert, P., 1986, 'The Medieval Monastic Garden', in Macdougall, pp. 23–54.

Miller, N.F., and Gleason, K.L. (eds), 1994, *The Archaeology of Garden and Field*, Philadelphia.

Miller, N.F., and Gleason, K.L., 1994, 'Fertilizer in the Identification and Analysis of Cultivated Soil', in Miller and Gleason, pp. 25–43.

Miller, S.G., 1977, 'Excavations at Nemea 1976', *Hesperia* 46, pp. 1–26.

Miller, S.G., 1978, 'Excavations at Nemea 1977', *Hesperia* 47, pp. 58–88.

Moynihan, E.B., 1979, *Paradise as a Garden in Persia and Mughal India*, New York.

Murphy, P., 1984, 'Carbonised fruits from Building 5', in Crummy, p. 40.

Murphy, P., 1992, 'Environmental Studies: Culver Street', in P. Crummy, *Excavations at Culver Street, the Gilberd School, and other sites in Colchester 1971–85*. Colchester Archaeological Reports 6, Colchester, pp. 273–87.

Murphy, P., and Scaife, R.G., 1991, 'The environmental archaeology of gardens', in Brown, pp. 83–99.

Murray, M.A., 2000, 'Fruits, vegetables, pulses and condiments', in Nicholson and Shaw, pp. 609–55.

Necipoğlu, G., 1991, *Architecture, Ceremonial and Power at the Topkapı Palace in the Fifteenth and Sixteenth Centuries*, Cambridge (Massachusetts).

Neudecker, R., 1998, *Die Skulpturenausstattung römischer Villen in Italien*, Mainz.

Nicholson, P.T., and Shaw, I., 2000, *Ancient Egyptian Materials and Technology*, Cambridge.

Noort, E., 1999, 'Gan-Eden in the Context of the Mythology of the Hebrew Bible', in Luttikhuizen, pp. 21–36.

Pailler, J.-M., 1987, 'Montmaurin: A Garden Villa', in Macdougall, pp. 205–11.

Parker, R.A., 1941, 'A late Demotic gardening agreement', *Journal of Egyptian Archaeology* 26, pp. 84–113.

Peet, T.E., and Woolley, C.L., 1923, *The City of Akhenaten, Part I. Excavations of 1921 and 1922 at El-Amarneh*, London.

Piccirillo, M., 1993, *The Mosaics of Jordan*, Amman.

Pinder-Wilson, R., 1976, 'The Persian Garden: *Bagh and Chahar Bagh*', in Macdougall and Ettinghausen, pp. 69–86.

Price, E., 2000, Frocester, *A Romano-British Settlement, its Antecedents and Successors*, Gloucester and District Archaeological Research Group.

Price, L., 1982, *The Plan of St Gall in Brief*, Berkeley.

Pritchard, J.B., 1955, *Ancient Near Eastern Texts Relating to the Old Testament*, Princeton.

Pritchard, J.B., 1969, *The Ancient Near East. Supplement to Texts and Pictures Relating to the Old Testament*, Princeton.

Renfrew, J.M., 1985, 'Preliminary report on the botanical remains', in Kemp, pp. 175–90.

Richardson, L., 1992, *A New Topographical Dictionary of Ancient Rome*, Baltimore.

Roebuck, C., 1951, *The Asklepeion and Lerna. Corinth*, Vol. 14, Princeton.

Rostovtzeff, M., 1922, *A Large Estate in Egypt in the Third Century BC*, Rome.

Ruggles, D.F., 2000, *Gardens, Landscapes and Vision in the Palaces of Islamic Spain*, University Park (Pennsylvania).

Saleh, A.-A., 1972, 'Some problems relating to the Pwenet reliefs at Deir-el Bahari', *Journal of Egyptian Archaeology* 58, pp. 140–58.

Schulz, R., and Seidel, M. (eds), 1998, *Egypt. The World of the Pharaohs*, Cologne.

Schulz, R., and Sourouzian, H., 1998, 'The Temples – Royal Gods and Divine Kings', in Schulz and Seidel, pp. 153–215.

Secchi, G.P., 1843, *Monumenti inediti d'un antico sepolcro di famiglia greca scoperto in Roma su la Via Latina*, Rome.

Simpson, W.K., 1973, *The Literature of Ancient Egypt. An Anthology of Stories, Instructions and Poetry*, New Haven.

Singer, C., 1927, 'The Herbal in Antiquity', *Journal of Hellenic Studies* 47, pp. 1–52.

Soren, D., 1987, *The Sanctuary of Apollo Hylates*, Tucson.

Stauffer, A., 1991, *Textiles d'Egypte de la collection Bouvier*, Bern.

Stronach, D., 1989, 'The royal garden at Pasargadae. Evolution and legacy'. *Archeologia Iranica et Orientalis. Miscellanea in Honorem Louis Vanden Berghe*, Ghent, pp. 475–502.

Stronach, D., 1994, 'Parterres and stone watercourses at Pasargadae: notes on the Achaemenid contribution to garden design', *Journal of Garden History* 14.1, pp. 3–12.

Tabbaa, Y., 1992, 'The Medieval Islamic Garden: Typology and Hydraulics', in Hunt, pp. 303–30.

Thomas, E.B., 1964, *Römische Villen in Pannonien*, Budapest.

Thompson, D.B., 1937, 'The garden of Hephaistos', *Hesperia* 6, pp. 396–425.

Tomei, M.A., 1992, 'Nota sui giardini antichi del Palatino', *Mélanges de l'Ecole française à Rome* 104, pp. 917–51.

Toynbee, J.M.C., 1971, *Death and Burial in the Roman World*, Baltimore.

Vercoutter, J., 1965, *Fouilles de Mirgissa*. Bulletin de la Société Française d'Egyptologie 43, Paris.

Villedieu, F. (ed.), 2001, *Il Giardino dei Cesari*, Rome.

Villedieu, R., 1992, 'Le Palatin (Vigna Barberini)', *Mélanges de l'Ecole française à Rome* 104, pp. 465–79.

Wendrich, W., 1990, 'Mats, baskets and plastic bags', *Egyptian Archaeology* 1993/3, pp. 39–40.

Wente, E.F., 1990, *Letters from Ancient Egypt*, Atlanta.

Wescoat Jr., J.L., 1990, 'Gardens of invention and exile: The precarious context of Mughal garden design during the reign of Humayun (1530–1556)', *Journal of Garden History* 10.2, pp. 106–16.

White, K.D., 1977, *Agricultural Implements of the Roman World*, Cambridge.

Wilber, D.N., 1979, *Persian Gardens and Garden Pavilions*, Washington.

Wilkinson, A., 1998, *The Garden in Ancient Egypt*, London.

Willcox, G.H., 1977, 'Exotic plants from Roman waterlogged sites in London', *Journal of Archaeological Science* 4, pp. 269–82.

Willeitner, J., 1998, 'Tomb and Burial Customs after Alexander the Great', in Schulz and Seidel, pp. 313–21.

Wiseman, D.J., 1983, 'Mesopotamian Gardens', *Anatolian Studies* 33, pp. 137–44.

Zanker, P., 1979, 'Die Villa als Vorbild des späten pompejanischen Wohngeschmackes', *Jahrbuch des Deutschen Archäologischen Instituts* 94, pp. 460–523.

Zeepvat, R.J., 1991, 'Roman gardens in Britain', in Brown, pp. 53–9.

Zimmer, G., 1982, *Römische Berufsdarstellungen*. Archäologische Forschungen 12, Berlin.

Zohary, D., and Hopf, M., 1993, *Domestication of Plants in the Old World*, Oxford.

Index

Figure numbers are in bold

Photographic acknowledgements

Every attempt has been made to trace accurate ownership of copyrighted visual material in this book. Errors and omissions will be corrected in subsequent editions provided notification is sent to the publisher.

Repeated floral motif reproduced from *Authentic Victorian Decoration and Ornamentation in Full Color* by Christopher Dresser (Dover Publications, Inc., New York).

Frontispiece Bibliothèque nationale de France; **1** The British Museum; **2** Maureen Carroll; **3** The British Museum; **4** Abdel Ghaffar Shedid, Munich; **5** Studium Biblicum Franciscanum Archive, Jerusalem; **6** Rheinisches Landesmuseum, Bonn, H. Lilienthal; **7** Maureen Carroll; **8** The British Museum; **9** German Archaeological Institute, Cairo, D. Johannes; **10** American School of Classical Studies at Athens, Agora Excavations; **11** Andrea Jemolo; **12** Jean Vercoutter/Institut de Papyrologie, Lille; **13** Barry Kemp; **14** The Metropolitan Museum of Art, New York, Rogers Fund and Edward S. Harkness Gift, 1920. (20.3.13) Photograph © 1992 The Metropolitan Museum of Art; **15** Courtesy of the Egypt Exploration Society; **16** The British Museum; **17** After M. Carroll-Spillecke 1989; **18** After M. Carroll-Spillecke 1989; **19** After M. Carroll-Spillecke 1989; **20** Maureen Carroll; **21** Maureen Carroll; **22** Maureen Carroll; **23** Maureen Carroll; **24** Maureen Carroll; **25** A. Klynne and P. Liljenstolpe 2000; **26** Barbara Ottaway; **27** Fishbourne Roman Palace; **28** After M. Carroll 2001; **29** Keith Branigan; **30** The British Museum; **31** The British Museum; **32** Lotos-Film, Kaufbeuren; **33** The British Museum; **34** Graham Harrison; **35** Studium Biblicum Franciscanum Archive, Jerusalem; **36** David Stronach; **37** Scala; **38** © The J. Paul Getty Museum; **39** Maureen Carroll; **40** Frank Brown; **41** The British Museum; **42** AKG London/P. Connolly and H. Dodge 1998; **43** Sheffield Library for Aerial Photography; **44** Maureen Carroll; **45** John Williams; **46** Nationalbibliothek, Vienna; **47** Courtesy of the Egypt Exploration Society; **48** After J.-C. Margueron 1992; **49** Department of Antiquities, Cyprus; **50** Vassos Karageorghis; **51** National Archaeological Museum, Athens; **52** Glyptothek, Munich; **53** After W. Jashemski 1979; **54** Stanley Jashemski; **55** Nationalbibliothek, Vienna/Bildarchive d. oNB, Wien; **56** The British Museum; **57** The Egyptian Museum, Cairo; **58** A.A. Van Heyden, Naarden; **59** Maureen Carroll; **60** Ekdotike Athenon; **61** After Jocelyn Toynbee 1971; **62** After W. Jashemski 1970/71; **63** Egyptian Expedition of the Metropolitan Museum of Art, Rogers Fund, 1930. (30.4.115) Photograph © 1978 The Metropolitan Museum of Art; **64** Nationalbibliothek, Vienna; **65** After W. Jashemski and E. Ricotti 1992; **66** Maureen Carroll; **67** Michael Holford; **68** American School of Classical Studies at Athens: Agora Excavations; **69** Nationalbibliothek, Vienna; **70** Maureen Carroll; **71** After Y. Barat and D. Morize 1999; **72** Robert Lloyd; **73** American School of Classical Studies at Athens: Agora Excavations; **74** Musée des Antiquités Nationales, St-Germain-en-Laye/© Photo RMN – Jean Schormans; **75** Accademia Italiana, London/Bridgeman Art Library; **76** The British Museum; **77** British Library; **78** Maureen Carroll; **79** Maureen Carroll; **80** Colchester Archaeological Trust; **81** Nationalbibliothek, Vienna; **82** The British Museum; **83** Canali; **84** Maureen Carroll; **85** © Musée d'art et histoire, Fribourg (Switzerland). Photo: Hans Kobi, Munchenbuchsee (Switzerland); **86** Maureen Carroll; **87** Musée des Antiquités Nationales, St-Germain-en-Laye/© Photo RMN – Jean Schormans; **88** Edward Price; **89** The British Museum; **90** British Library; **91** Courtesy of the Trustees of the V&A; **92** Editions d'Art Albert Skira, Geneva; **93** Topkapi Museum, Istanbul; **94** Bodleian Library, University of Oxford; **pp. 136–7** Maureen Carroll